Exploring Algebra 1

with

ERIC KAMISCHKE

LARRY COPES

ROSS ISENEGGER

Project Editor:	Joan Lewis
Editorial Assistants:	Christa Edwards, Aneesa Davenport, Elizabeth Ball
Reviewer:	Corey Andreason
Accuracy Checker:	Dudley Brooks
Field Testers:	Cynthia Beals, Barbara J. Lamberski, John Olive
Production Editors:	Angela Chen, Donna Crossman
Copyeditor:	Tara S. Joffe
Production Coordinator:	Jennifer Young
Vice President, Editorial and Production:	Casey FitzSimons
Text Designer:	Marilyn Perry
Compositor:	Graphic World
Cover Designer:	Jensen Barnes
Prepress and Printer:	Lightning Source, Inc.
Publisher:	Steven Rasmussen

Key Curriculum Press
1150 65th Street
Emeryville, CA 94608
510-595-7000
editorial@keypress.com
www.keypress.com

Printed in the United States of America

10 9 8 7 6 5 4 3 2 15 14 13 12 11 ISBN 13: 978-1-55953-801-5

Contents

5: Transforming Functions

6: Investigating Higher-Degree Polynomials

7: Simulating Probability

Downloading Fathom Documents

Getting Started

All Fathom documents for *Exploring Algebra 1 with Fathom* are available online for download.

- Go to www.keypress.com/ExpAlg1Fathom. Click "Download Activity Files."

The downloadable folder contains all of the documents you need for this book, organized by chapter and activity. The documents require Fathom Version 2 software to open. Go to www.keypress.com/ftm/order to purchase Fathom, or download a trial version from www.keypress.com/ftm/download.

Types of Documents

Activity documents can include simulations setups, sample data, or presentation documents to use for presentation. You can display files with *Present* in their filenames, using a computer and a display device, to demonstrate mathematical relationships and generate whole-class discussions. In the Activity Notes, you'll find information on using the documents.

Using the Activities

Data tie algebra to the real world. Rate and starting value can be modeled by slope and y-intercept; different ways of looking at ratios can be shown on ribbon charts or bar graphs; feasible regions can be modeled by systems of equations or inequalities; animation can be programmed as algebraic transformations; business decisions about sales price and profits can be based on quadratic models; probability can be simulated to give experimental results quickly. In each of these situations and many more, algebraic models and their graphical representation help students answer questions about things they are familiar with outside the mathematics classroom.

Using these activities, students will deepen their understanding of the underlying mathematical concepts as they develop lifelong habits of learning through asking questions, such as: What happens if . . . ? Under what conditions . . . ? or Is there a relationship . . . ? In the process, they will learn to use Fathom, which is an ideal dynamic environment for exploring questions like these. Fathom allows students to

- plot functions on a scatter plot to find a line or curve of fit
- drag data to see the effect one value has on the statistics that describe the whole data set
- build simulations and watch sampling

These activities cover the major topics of Algebra 1 and relate to a variety of interests. Think of them as starting points. Refer to them for ideas on how to have students use Fathom to solve problems from your curriculum. Even better, encourage students to find their own data and to use Fathom while pursuing questions that interest them. For many of the activities, students are given a chance to choose and explore data on a country, part of the United States, or a gender. The extensions recommend directions you could suggest for further student exploration. The deepest learning takes place when students are answering questions they themselves have asked.

When to Use an Activity

The activities are designed to be independent of each other. If students are unfamiliar with Fathom Dynamic Data software, you might want to start with the early activities in Unit 1 or Unit 2, which assume very little Fathom experience. Or you might use one of these early activities as a presentation of the kinds of questions Fathom can answer and the methods used.

The audience for these activities is beginning algebra students. However, because learning is spiral and boundaries between courses are blurry, many of the activities can also be used in pre-algebra, advanced algebra, or integrated courses. The student audience listed in the Activity Notes will help you determine the activities that will be appropriate for your students.

The chart of activities on page ix, unit and activity titles, and objectives listed in the Activity Notes will all help you choose appropriate activities for particular topics.

Activity Settings

The activities in this book are flexible enough that they can be used in a variety of instructional settings.

- **Paired/Individual Activity:** An ideal classroom situation for exploring with dynamic software is a computer for every two or three students. As students talk together about what they are doing, they benefit from each other's strengths and unique experiences. Be aware that students with certain kinds of red-green color blindness may have trouble seeing red highlighting. Make sure those students work cooperatively with students who are not color blind. You might notice that some students are "keyboard hogs." Halfway through the activity, you should direct students to switch who is handling the keyboard.

As you circulate among student groups, you may find yourself answering questions about the software. As you identify students who become comfortable with Fathom very quickly, enlist their assistance in helping other students. You want to be able to spend your time asking leading questions, such as: What have you tried? Why did you choose that representation? Could you compare . . . ? Is there another way to . . . ? You also want to look for unusual displays or approaches that students can share with the class.

It's best to give students or groups a chance to tell the class about what they have done and learned. Students benefit from seeing others' discoveries and from explaining their own. The sharing becomes even more interesting when students have pursued their own questions or the extensions in the Activity Notes.

- **Paired/Individual Exploration:** A few activities are written as explorations. Students are given a situation to explore and a question to answer, but they are not told which Fathom objects to use or how to use them. These explorations are for students who are familiar with Fathom and who are experienced investigators. Completing these explorations can give students the kind of experience they will need to use Fathom to explore their own questions about data and mathematics.
- **For a Presentation:** At times, you might want to use Fathom to demonstrate a concept. You can adapt any of the activities. Some Activity Notes have suggestions for using the activity as a whole-class presentation, for which files named with the word *Present* are appropriate. Before using a Fathom document for a presentation, you might set the Fathom **Preferences** at a larger font size for easier viewing.

When you present, involve students. Ask a student to run the computer, involve the class in deciding what to graph, and ask questions that promote student involvement. Before you create a graph, show summary data, or run a simulation, ask questions, such as: What do you predict the graph will look like? Which do you think will be larger? What do you think will happen?

Fathom Skills

Activities 1.1–1.3 and 2.1–2.3 can be used to introduce basic Fathom knowledge. The other activities assume students can open new or existing Fathom files, create a case table, add attributes and cases, graph attributes of a data collection, and use sliders, though many activities include margin notes reminding students of these basic tools. Later activities gradually introduce the skills of adding lines to graphs, building and reading summary tables, using formulas in attributes and sliders, and taking samples and measures. By the last units, students are building simulations.

Activity Notes indicate Fathom skills assumed and Fathom skills introduced. Use this information to anticipate how much Fathom coaching students may need. Be prepared to be surprised at how easily students figure out this new technology. Once they understand the basic structure of a collection (often most easily understood as the cases listed in a table) and know they can represent that collection as a dynamic graph (controlled by sliders), they'll find ways to do what they want to do.

One Fathom skill is arranging collection, table, graph, summary, and slider objects in the Fathom window. Although viewing a full table of data helps students see what they are working with, tables may monopolize a large part of the window. Once students understand the cases and attributes, the table can be shrunk to allow room for other objects. You and your students will gradually learn how to set up the Fathom window so that objects are organized and available. For example, you'll find that it is best to arrange sliders near the displays they affect.

Enjoy exploring the activities, sharing the mathematics, and developing technology experience so that Fathom becomes a dynamic tool students use to answer questions about mathematical or statistical data. As their guide, you will ask questions that help them think about what they are doing and make the connections that deepen their mathematical understanding. You will also help students find answers to questions about using the software. Keep the focus on student questions and their ability to pursue them, aided by their peers and Fathom.

Chart of Activities

Use this chart to locate activities that are appropriate for various lessons in your curriculum.

Activity	Pre-algebra	Algebra 1	Algebra 2	Average Time (minutes)	Paired/Individual	Whole-Class	Exploration	Fathom Document	Presenter Document	Description
1: Building on Pre-algebra										
Measures of Center—Aircraft Efficiency	○	○		50–70	○	○		○		Investigate mean and median and look at attributes that contribute to efficiency.
Spread and Shape—Dice	○	○		45–60	○	○		○		Explore spread and shape of data generated from rolling one or two dice.
Box Plots—Insurance	○	○		35–50	○	○		○		Graph and compare life expectancy data (male, female) within countries and between countries.
Box Plots—Think Box	○	○		40–60	○	○		○	○	Drag data points to see how changes affect the box plot.
Histograms—Ontario Communities	○	○		20–35	○			○		Interpret data on geographic location and income; explore how changes in bin width affect histograms.
Ratios—Surveys	○	○		25–40	○					Examine ways to represent ratios; use different ratios to solve proportions for unknowns.
Weighted Average—Swimmers	○	○		30–45	○			○		Explore ratios used to calculate weighted average; use sliders to determine weights.
Proportions—Veterans	○	○		20–30	○			○		Set up and solve proportions using census microdata and compare statistics between U.S. states.
Dimensional Analysis—Fastest Animals	○	○		30–40	○			○		Use unit ratios to convert units, make connections between ratios and measurement, and validate conversion factors.
Proportions—Squirrel Population		○		30–45	○	○		○		Simulate several capture-recapture scenarios; set up and solve proportions.
2: Exploring Linear Equations										
Slope—Runners	○	○		25–35	○	○		○		Understand the slope of a line as a constant rate; calculate the slope of a line by dividing vertical change by horizontal change.
Lines of Fit—Women's High Jump	○	○		20–25	○			○		Use movable lines to identify a line of fit between two points and find its slope.
Intercept Form—Hot Dogs		○		30–40	○					Understand slope as the rate of change and recognize it as the coefficient in the linear equation.
Point-Slope Form—Men's High Jump		○		20–30	○	○		○	○	Use a movable line and new attributes to write the equation of a line in point-slope form.
Point-Slope Form—Life Expectancy		○		35–50	○	○	○	○		Develop and use the point-slope form for the equation of a line.
Linear Modeling—Dissolved Oxygen		○		25–35	○			○		Understand the need to develop a method for finding lines of fit; use quartile points to find a line of fit.

Activity	Pre-algebra	Algebra 1	Algebra 2	Average Time (minutes)	Paired/Individual	Whole-Class	Exploration	Fathom Document	Presenter Document	Description
2: Exploring Linear Equations continued										
Equation Solving—Cryptography	○	○		30–40	○					Practice undoing a linear expression as a way to solve an equation.
3: Solving Inequalities and Systems										
Inequalities—Strange Diet		○		50–60	○			○		Translate inequality statements from context to symbols; solve inequalities; interpret a solution in context.
Absolute Value—Radio Contact		○		25–35	○			○		Solve inequalities with absolute values; understand solution sets and the logical meanings of *and* and *or*.
Two-Variable Inequalities—The Quest		○		30–40	○	○		○	○	Solve linear inequalities with two variables; graph inequalities using shading.
Linear Systems—The Road Trip	○	○		15–25	○	○		○		Use graphs of linear systems of equations to model and solve real-world problems.
Systems Solving—High Jump Records		○		30–40	○			○		Use substitution to solve a system of linear equations modeling a real-world problem; interpret results.
Elimination—Package Charges		○		40–50	○	○		○	○	Discover that the sum of two linear equations in a system is a linear equation whose solutions include the solution of the system; visualize what the elimination method represents graphically.
4: Exploring Exponential Equations										
Exponential Growth—Interest		○		35–50	○	○				Learn that a linear equation models simple interest and an exponential equation models compound interest.
Exponential Relationships—Population Growth		○	○	45–60	○			○		Use an exponential equation to model population growth and learn how the base is a constant ratio of consecutive values.
Inverse Variation—Boyle's Law		○	○	30–40	○			○		Model inversely proportional quantities and investigate the change in one variable as the other doubles.
Exponents—Moore's Law		○	○	50–75	○	○		○		See the same data modeled two ways and discover the multiplication property of exponents.
Power Properties—Base *e*		○	○	45–70	○					See that continuous compounding of an investment at rate r for one year is modeled by the exponential equation $y = e^r$.
Power Properties—Radiation		○	○	45–70	○			○		Investigate an exponential decay model and see $(c^a)^x = c^{(ax)}$.
5: Transforming Functions										
Functions—Model Rockets		○		20–30	○	○				Understand that a function has no more than one output for each input but may have more than one input with the same output; learn and apply *independent* and *dependent variable, domain, range.*
Data Transformation—Quiz Scores		○	○	25–35	○			○		See how the position and spread of a data set are changed by shifts and stretches; review *median.*

Activity	Pre-algebra	Algebra 1	Algebra 2	Average Time (minutes)	Paired/Individual	Whole-Class	Exploration	Fathom Document	Presenter Document	Description
5: Transforming Functions continued										
Transformations—Animation		○	○	45–60	○			○		Explore two-directional translation and dilation of line plots.
Line Transformations—Elevator		○		30–45	○					Investigate vertical and horizontal translations of lines and relate transformations of data to transformations of functions.
Parabola Transformations—Handshakes		○	○	30–45	○	○				Learn to modify a quadratic equation in order to translate its graph.
Parabola Transformations—Book Sales		○	○	20–30	○		○	○		Model quadratic data; deepen understanding of transformations; explore dilations of functions.
6: Investigating Higher-Degree Polynomials										
General Form Quadratic—Escape Ramp		○	○	25–40	○	○	○	○		Solve a real-world problem and dynamically explore how a graph of the general form of a quadratic is determined by its coefficients.
Factored Form Quadratic—Gravity		○	○	25–35	○					See relationships among the coefficients of a quadratic equation in general form, the factored form of the equation, and the function's zeros; use sliders to discover how the factored form relates to the graph.
Vertex Form Quadratic—Protecting Wildflowers		○	○	30–45	○	○	○	○		Create a strategy for modeling quadratic data (based on a vertex-dilation equation); use sliders and units to discover the roles of values used in the vertex form of the quadratic.
The Quadratic Formula		○	○	25–35	○		○			Explore the roles of coefficients of quadratic functions in locations of zeros, as given by the quadratic formula, and in the location of x-intercepts of the function's graph.
Parabola—Solar Oven		○	○	25–35	○	○				Use a quadratic equation to model a solar oven; discover relationships between the quadratic formula and the graph.
Binomial Products—Sales and Profits		○	○	30–40	○	○		○		Explore how linear models can give both graphical and symbolic information about the quadratic model that is their product.
Binomial Expansion—Flipping Coins		○	○	15–25	○			○		Explore how patterns in binomial expansions relate to flipping coins.
Common Factor—Acceleration		○	○	15–25	○			○		Use factoring to find a quadratic model for real-world data involving motion on an incline.
Polynomial Factoring—Maximum Area		○	○	20–35	○		○	○		Use factoring as a process to model polynomials; learn about relationships among intercepts, zeros, and factors.
7: Simulating Probability										
Probability—Dice Games		○	○	50–60	○	○		○	○	Use Fathom to simulate a game based on a historical question that led to probability as a branch of mathematics.

Activity	Pre-algebra	Algebra 1	Algebra 2	Average Time (minutes)	Paired/Individual	Whole-Class	Exploration	Fathom Document	Presenter Document	Description
7: Simulating Probability continued										
Probability—Euchre Deck		○	○	30–40	○	○		○	○	Create simulations to answer probability questions; contrast sampling with and without replacement; understand the difference between independent and dependent events.
Simulation—Stick or Switch		○	○	25–40	○			○		Make sense of probability; learn about conditional probability; create probability simulations.
Geometry by Probability—Monte Carlo Method		○	○	25–35	○					Use a probability simulation to estimate area.

1

Building on Pre-algebra

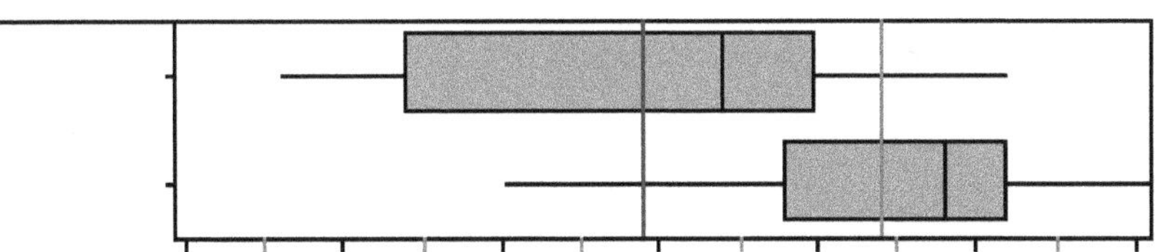

Measures of Center—Aircraft Efficiency

The airline business has gone through turmoil recently, with bankruptcies and reorganizations. To keep costs down and remain profitable, it is important for airlines to use efficient aircraft. In this exploration, you'll examine some data and make recommendations about the most efficient aircraft to use.

In describing the airline data, you'll probably want to mention values that are typical, or in the middle of the data. You often hear statements about typical values, such as *Most men have shoe sizes 9 to 11* or *The average university student gains 15 pounds in the freshman year* or even *The median score on the Algebra 1 exam was 72%.* These statements refer to the *central tendency* of data sets.

Q1 Two measures of central tendency are the mean and median. How do you calculate the *mean* (or *average*) of 36 numbers? How do you calculate the *median* (or middle number) of 36 numbers?

Q2 Can the mean and median of 36 numbers be different? Can they be the same? Use examples to justify your answers.

INVESTIGATE

You have been asked to recommend the best plane for your company to use on a new route. The data you're given about airplanes are contained in a Fathom document.

A shorter version of this set of instructions is Choose **File | Open | Aircraft.ftm.**

1. Start the Fathom program. Go to the **File** menu and choose **Open.** Browse until you find the document called **Aircraft.ftm.**

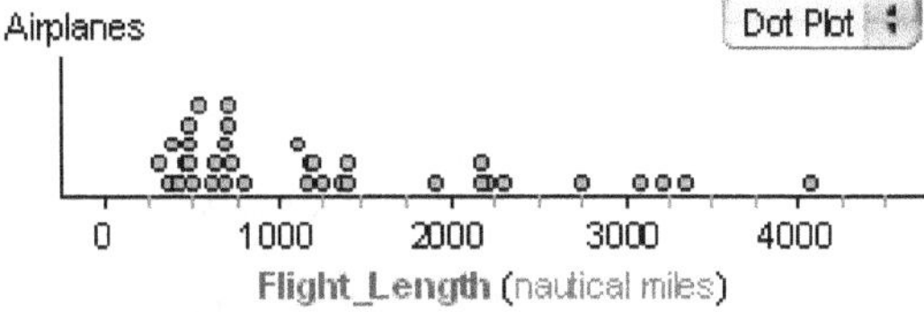

On the screen, you will see three objects: a box of balls, called a *collection;* a graph, called a *dot plot;* and a table, called a *case table.* Each ball represents one airplane, or one *case* in the data collection. The planes are also listed as cases in the table. You can see the *attributes,* or the information available for each plane, listed across the top of the case table. Notice that five of the plane attributes have units, which are displayed in green on the screen. The dot plot shows one dot for each plane, representing that plane's flight length.

You can enlarge the collection or the case table by dragging the edge.

2. Make sure you understand what each attribute means and how the collection and the case table are related. Browse through the case table.

Q3 For how many planes does the case table contain data? Which plane has the longest range (*Flight_Length*)? Which airplane is the slowest (has the smallest value for *Speed*)?

You can also use the graph to answer Q3.

3. Click the rightmost dot on the dot plot. Notice that the plane represented by this dot is now highlighted in the case table. While the cursor is over the dot, Fathom displays information about this airplane in the status bar at the bottom left of the screen. The rightmost dot corresponds to the plane with the longest flight length. The values displayed at the bottom of the screen should agree with your answer to Q3.

Q4 Look at the 36 dots in the plot and think about their flight lengths. Estimate what you think a typical value for *Flight_Length* would be.

4. You can use Fathom to calculate measures of central tendency. Click on the dot plot to select it. Go to the **Graph** menu and choose **Plot Value.** You'll see a new window, called the *formula editor.*

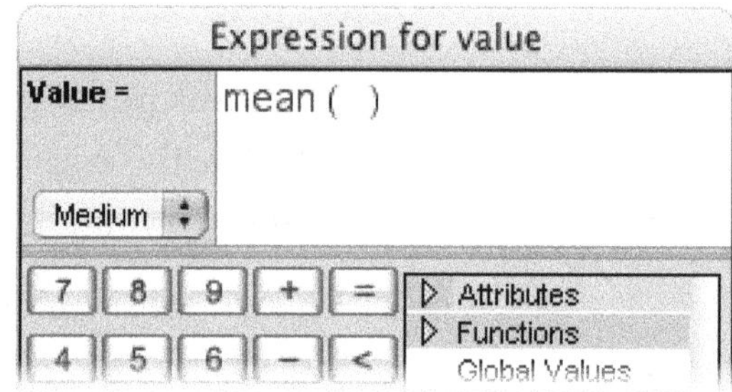

Type mean( ) (with the parentheses) and click **OK.** Fathom will calculate the mean of the 36 flight lengths, display the value at the bottom of the graph, and draw a vertical line corresponding to the mean's value on the graph. Again choose **Plot Value,** and plot the value median( ).

Q5 Record the mean and median of these data and compare them with the estimate you made in Q4.

Q6 Why do you think the mean is greater than the median for this data set?

Q7 Does the mean or the median give a better indication of the flight length of the typical aircraft? Why?

Q8 Does your answer to Q7 help narrow down your recommendation for which plane an airline should use? Explain your answer.

5. Drag the graph icon from the object shelf to some white space in the Fathom window. Now drag the attribute *Operating_Cost* from the case table onto this new graph and drop it where it says **Drop Attribute Here.** This will create a dot plot that shows operating cost per hour.

Q9 Click on the dots on the new graph to determine which planes have the lowest and highest operating costs. Also find the mean and median of the operating costs.

Q10 Based on your answers to Q9, what recommendations would you make about which plane the airline should use?

Q11 Experiment with other attributes. Which attributes are most important when considering aircraft efficiency? What combination of attributes might make a good efficiency rating? You will develop a formula for efficiency involving several attributes when you complete the Explore More.

You may wonder how modifying some of the airplanes changes the mean or median. Could you enlarge the fuel tank to give it a longer flight length? How would that affect the mean and median flight length? With Fathom, you can change the values in the data to see what effect this has on the mean and the median. For any attribute you are considering, drag the rightmost dot to the left. The corresponding value in the table will change.

Q12 Explore to answer these questions. Use complete sentences to describe how changing values in the data affects the mean and the median.

- How far do you have to move the dot to make the mean change?
- How far do you have to move the dot to make the median change?
- How many dots did you have to move to make the mean equal to the median?

If you want to undo any change in Fathom, choose **Edit | Undo.** You can undo repeatedly until the data set is back to its initial values. Another way to get back to the original data is to select the collection (the box of golden balls) and choose **File | Revert Collection.**

Q13 Review your answer to Q2. Explain the conditions that make the mean equal to the median.

EXPLORE MORE

You might use the fact that 1 knot (kt) = 1.152 miles per hour (mph). Or you can change the units from kt to mph in the case table, and Fathom will do the conversion for you.

A December 5, 2005, article in *Newsweek* about the new Boeing 777 compared the plane's efficiency to that of the Airbus A380, saying, "The Boeing burns 24 percent less fuel per seat mile." Create a formula for the aircraft efficiency that will take into account the cost per seat mile. Your formula could take into account other attributes as well.

Scroll to the far right of the table. You will see an attribute called *<new>*; click on it and type efficiency. Double-click the gray field below *efficiency.* Enter your formula for *efficiency* and click **OK.** Create the dot plot for your new attribute. What are its units? Are you looking for the plane with the largest or smallest value? Which plane do you think is the most cost efficient to fly? Does your answer agree with the answers of others in your class? What are the mean and median of your measure of efficiency?

Measures of Center—Aircraft Efficiency

Objective: Students will investigate the measures mean and median, using the dynamic aspect of Fathom to observe the effect that changing one data value has on the mean and median. Students will consider the ability of the mean and the median to measure central tendency.

Student Audience: Pre-algebra, Algebra 1

Activity Time: 50–70 minutes

Setting: Paired/Individual Activity or Whole-Class Presentation (use **Aircraft.ftm** for either setting)

Mathematics Prerequisites: Students can calculate the mean and median of a data set.

Fathom Prerequisites: Students can open Fathom sample documents.

Fathom Skills: Students will interpret data in case tables and graphs, create a dot plot, use the formula editor to plot a computed value on a graph, change the value of an attribute for one case by dragging a dot on the graph, and use **Undo** or **Revert Collection** to return a collection to its saved state.

Notes: This activity requires a long class period (or two periods) for students to complete the activity and define an efficient aircraft in the Explore More. If possible, give students time to explore Q11 and the Explore More. Take time for students to share their answers with the class and to explain the thinking behind their formulas.

For a Presentation: To use this activity to introduce Fathom to a class, ask the questions in the investigation as you show the features of Fathom that are introduced. Questions such as Q4, Q6, and Q10 can solicit several answers and begin good discussions. After the demonstration and discussion, you might have student groups work on the Explore More. If groups do not have computer access, you could distribute copies of the table and ask groups to devise formulas, then come up and add their attribute with its formula to the demonstration computer.

Q1 The mean is the sum of all 36 values divided by 36. The median of a data set with an odd number of cases, ordered from least to greatest or greatest to least, is the middle number. The median of an ordered data set with an even number of cases is the mean of the two middle values. When the data are sorted from smallest to largest, the median of a data set with 36 values will be the mean of the 18th and 19th value.

Q2 It is quite likely that the mean and median of a data set are different, because they are calculated differently. It is possible to create a data set for which the mean and median happen to be the same (such as 1, 1, 1, 1, 1 or 1, 2, 3, . . ., 36).

INVESTIGATE

1. Many Fathom sample documents have comments related to their source and composition. Double-click on the collection and click the **Comments** panel tab to view this information.
2. Managing screen real estate is an important skill for new Fathom users, as the workspace tends to get cluttered with tables, graphs, and inspector windows. Resizing and positioning these objects will become natural after some practice.

Q3 The table records data for 36 planes. To find the plane with the given characteristics, you can scroll through the data carefully or sort the data. To find the plane with the longest range, select *Flight_Length* in the case table and choose **Table | Sort Descending.** The B747-400 has the longest range, at 4065 N.M. Sorting on *Speed* in ascending order reveals that the ERJ-135 is the slowest, at 328 kt.

3. "B747-400 (4065 N.M.)" is displayed at the bottom left.

Q4 The estimate for the typical value should be between 750 and 2000 N.M.

4. The formulas could be entered as mean(Flight_Length) and median(Flight_Length). (Fathom automatically supplies the closing parenthesis.)

Q5 The mean flight length is 1313.33 N.M., and the median is 949.5 N.M. The comparisons will vary.

Q6 The mean is larger because a couple planes have long flight lengths. This type of data set—with a tail on the right when graphed—is referred to as *skewed right.*

Q7 The median value better reflects where the middle of the majority of the data is located, but the mean takes into account the data spread out on the right. Together, both provide good information.

Q8 Answers will vary.

Q9 The operating costs are highest for the B747-100 and lowest for the ERJ-135. The dot plot for *Operating_Cost* is similar to the dot plot for *Flight_Length,* as they both have a clump of planes at the smaller values and a few planes spread out along the larger values. This spread results in a mean that is greater than the median. Reasonable student predictions should have the median and mean between 2000 and 4000, with the mean being the greater of the two.

Q10 The ERJ-135 has the lowest operating cost, \$650/h. However, it is the slowest plane, it only seats 37, and it has one of the smallest ranges. This makes it impractical for commercial use.

Q11 Answers will vary.

Q12 The mean changes when any value changes. The median changes if a changing value becomes the 18th or 19th largest (the only two values that determine the median) or if values are moved to the other side of the median, which would affect the values in the 18th and 19th places. For these data, because there are 36 cases, the mean changes 1/36th as much as the change in the case's value and in the same direction. Originally, the 18th and 19th values of *Flight_Length* are 795 N.M. and 1104 N.M., respectively, and the median is the mean of these two values. Dragging the largest value to the left eventually causes it to become the 19th largest and then the 18th largest (as it joins the smaller half of the data); if it gets smaller than 707 N.M., then 707 N.M. becomes the 18th and 795 N.M. the 19th. The median can take on any value between 751 N.M. (the mean of 707 N.M. and 795 N.M.) and 945.5 N.M. (the mean of 795 N.M. and 1104 N.M.). As the value moves in this range, the median changes half as fast as the value does.

Answers will vary for the number of dots that need to move for the mean to equal the median. For flight length, moving the five longest flight lengths to the left until just before the median makes the mean and median approximately equal.

Q13 Under certain circumstances, the mean and the median can be the same. The variation on one side of the median needs to be balanced by the variation on the other side; the data are approximately symmetrical, with no outliers on only one side.

EXPLORE MORE

Operating cost per seat gives efficiency in dollars per seat hour. To change to dollars per seat mile, divide by speed in miles per hour. According to this formula, the most efficient aircraft is the B737-800, with 149 seats, speed of 454 kt, operating cost of \$1665/h, and efficiency \$0.0185664/seat mile. Note that Fathom originally calculated the formula in dollars per [seat] meter. When you change the *m* to *mi* in the units, Fathom calculates the conversion.

Speed	Efficiency
mi/h	dollars/mi
	Operating_Cost / Seats / Speed

Some student equations will not work for planes that have no passenger seats. You might suggest they delete these cases when they look for efficiency formulas for passenger aircraft. Units on some formulas may be confusing; Fathom works in nautical miles and knots rather than miles per hour.

EXTENSION

Arthur Hailey wrote *Flight into Danger,* a screenplay that first aired in 1956. It launched his hugely successful career as an author, creating blockbuster movies and even spoofs of those movies. Research the impact of *Flight into Danger* and Hailey's work, and reflect on how society's mixed feelings toward technology still provide plots for today's popular books and movies.

Spread and Shape—Dice

People have created many games of chance with numbered cubes. To help you analyze the chance of winning these games, Fathom can simulate how dice and their sums behave. Although simulations of random chance will vary in the short term, they will show you what can happen in the long run. You will see that data sets representing the possible outcomes in two games can contain the same elements but have different chances of winning.

Q1 What are the mean and the median of the data set $\{1, 2, 3, 4, 5, 6\}$?

Q2 If you roll a fair die 600 times and record the face value (1, 2, 3, 4, 5, or 6) for each roll, roughly how many times would you expect to roll a 1? What would you expect the mean and median of the face values for 600 rolls to be?

INVESTIGATE

When you start Fathom, a blank document will open. You want to set up a collection that holds 600 simulated rolls of a die.

1. From the object shelf, drag a new collection (box of golden balls) to the window. Double-click on the picture of the box to bring up a new window labeled Inspect Collection 1.

If you see a Rename Collection dialog box, then you double-clicked on the collection's name rather than on its picture. You might want to give the collection a new name before you click again on the collection.

Attribute names can't have spaces. An underscore can be used in place of a space.

2. Data in a Fathom collection have one or more attributes. To define an attribute, click on <*new*> in the inspector's Attribute column. Then enter the name of the attribute. In this case, enter Roll_1_to_6.

To get actual numbers into that attribute, you'll use a formula to generate random integers from 1 to 6, representing the face values on a die.

3. Double-click opposite *Roll_1_to_6* in the Formula column. Fathom's formula editor will open. Type randomInteger(1, 6) and click **OK.** Move the inspector out of your way, but do not close it.

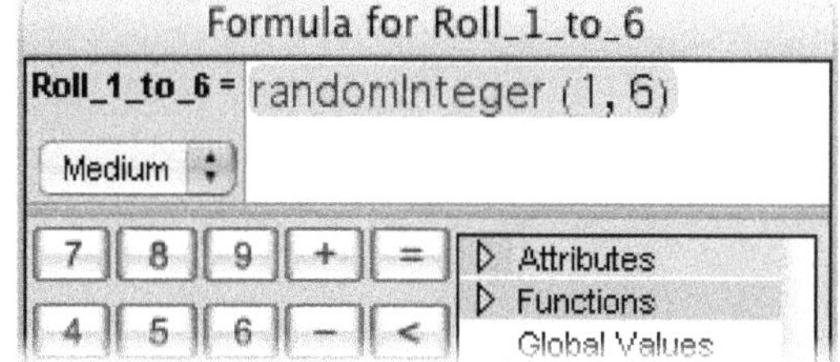

4. Click on the collection to select it. Choose **Collection | New Cases** and add 600 new cases.

5. Drag a graph from the shelf into your document. Drag the attribute *Roll_1_to_6* from the inspector to the graph's horizontal axis. With the graph selected, go to **Graph | Plot Value** and type the formula mean( ). Click **OK.** In the same way, plot the value for the median.

6. It may be difficult to count all those dots. A bar chart with the heights marked might be more useful. Drag down a new graph. Again drag *Roll_1_to_6* to the horizontal axis, but this time hold down the Shift key as you release the mouse button. You have a bar chart!

Q3 Do these graphs show the values you expected? Explain.

You have done one simulation of 600 cases. You can use Fathom to repeat the simulation many times.

The keyboard shortcut Ctrl+Y (Win) or ⌘+Y (Mac) will come in handy.

7. From the **Collection** menu, choose **Rerandomize.** A new random number will be calculated for each case, and the graphs will be updated. Rerandomize the collection repeatedly.

Q4 Describe the general shape shared by all the different graphs generated by rerandomizing. What values does the mean have? What values did you get for the median? What three values can the median have?

Finding the sum of the rolls is often important in games. A sum of two dice would be one of the values 2, 3, 4, 5, 6, 7, 8, 9, 10, 11, or 12.

Q5 What would you expect the mean and median of the sum to be?

You could simulate the set {2, 3, 4, 5, 6, 7, 8, 9, 10, 11, 12} using random numbers from 2 to 12.

8. Just as you did for rolling 1 to 6, create a new attribute called *Roll_2_to_12*, and enter the formula randomInteger(2, 12). Add several hundred new cases to simulate at least 1000 rolls.

9. Create a dot plot of *Roll_2_to_12* and plot the value(s) of the mean and median. Create its bar chart as well. Rerandomize the values in the collection repeatedly and observe the changes in the dot plot and bar chart.

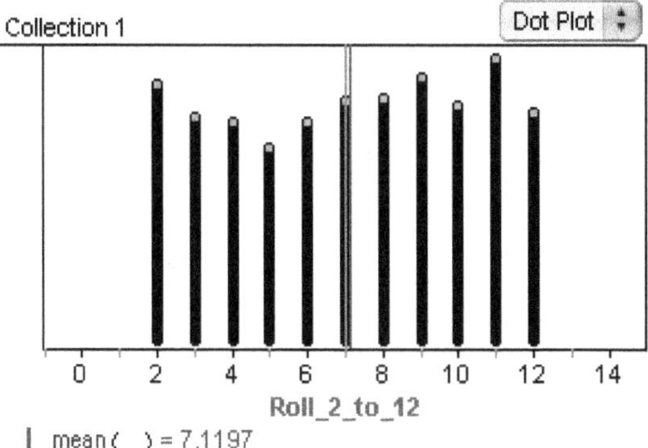

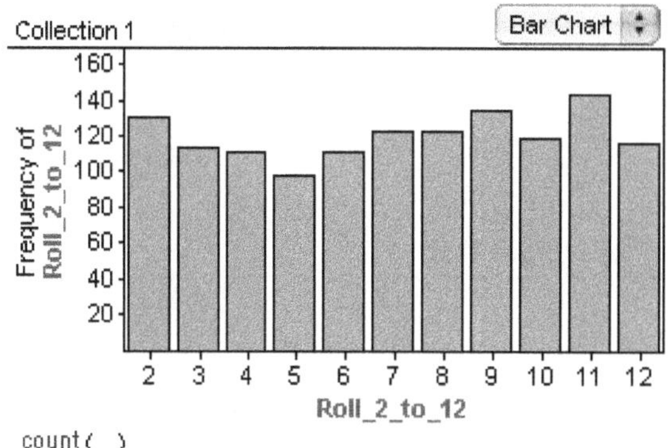

Q6 What are the mean and the median face values of these data sets of more than 1000 rolls? Does this agree with your answer to Q5?

Q7 The *range* of a data set is the difference between the maximum value and the minimum value. As you rerandomize, does the range change?

Q8 As you rerandomize, the mean changes. What was the smallest and largest value you observed for the mean?

Q9 How well do you think this formula simulates the sum of two dice?

Another way to simulate the sum would be to add the results for two individual dice.

10. Create a third attribute, *Add_Rolls.* Give it a formula that simulates two dice and adds their values.

Q10 What formula did you use?

11. Create both the dot plot (with the mean and median) and the bar chart, as you did for the previous two collections. Rerandomize repeatedly to explore the general features of this collection's dot plot and bar chart.

Q11 What are the mean, median, minimum value, maximum value, and range of *Add_Rolls?* Measures of data like these numbers are called *summary statistics,* because they can be used to summarize the behavior of a lot of data (such as more than 1000 rolls of a die). How do these summary statistics compare with those for randomly choosing a number between 2 and 12?

A summary statistic is useful in keeping track of thousands of values, but it can hide or ignore much of the information and thus can be misleading. For example, the summary statistics we've been looking at don't describe the shape of the data when graphed.

Q12 How does the graph of *Add_Rolls* compare with the graph of *Roll_2_to_12?* Which graph would you say is more spread out (less bunched up in the center)? How do the graphs differ in showing the chances to win with a given roll?

Q13 How well does *Add_Rolls* simulate finding the sum of two dice rolled randomly?

EXPLORE MORE

1. If you had three tetrahedral dice (4-sided), then the sum of the dice would always be between 3 (which is 1 + 1 + 1) and 12 (which is 4 + 4 + 4). How would the shape and spread of this distribution be different from that of a 6-sided die and that of the sum of two 6-sided dice? Write your prediction and then use Fathom to test it.
2. Plot the value s( ) on your dot plots of *Add_Rolls* and *Roll_2_to_12.* This is the spread measure called *standard deviation.* How does the value of *s* relate to how much the graph is spread out (that is, less bunched up in the center)?

Spread and Shape—Dice

Objective: Students will create bar graphs in Fathom to model dice games involving random chance; look at mean, median, maximum, minimum, and range; and interpret the results. Using Fathom to quickly create many simulations, students will see the importance of spread and shape when comparing distributions with the same center and range but with different shapes and spreads.

Student Audience: Pre-algebra, Algebra 1

Activity Time: 45–60 minutes

Setting: Paired/Individual Activity or Whole-Class Presentation (**Dice.ftm** is optional for either setting.)

Mathematics Prerequisites: Students understand the terms *mean, median, range, maximum,* and *minimum* for a data set.

Fathom Prerequisites: Students can start Fathom.

Fathom Skills: Students will learn to create a new collection and add attributes, create a formula involving the randomInteger function for an attribute, add cases to a collection, create dot plots and bar charts of numerical data, plot the value of statistical measures on a dot plot, and use **Rerandomize.**

Notes: This lesson is an excellent introduction to Fathom and its capacity to simulate probability events. Rather than starting with a table showing the cases in a collection, students start with a blank document and build a collection of dice rolls. Fathom's dynamic interface will demonstrate the variability that is present in any model involving randomness. The randomInteger function is used to simulate an equal-probability event such as rolling a die. Fathom can display the cases as dice; the document **Dice.ftm** includes the dice display and the first attribute, *Roll_1_to_6.* Starting with that document might help students see that a case is one roll of the die. Or you can show this document and rerandomize a couple of times so students see what is happening. Enlarge the collection's window so students can see the result of many of the 600 rolls. Then let groups work through the activity.

For a Presentation: You might use this presentation to introduce Fathom and show how it can be used to simulate a random event. As you facilitate the discussion, encourage several students to describe and explain what they see in Q3 and Q4. Get several opinions on Q5 and Q9 and several comparisons for Q12. Ask for predictions before you (or a student running the computer) create a graph or add the mean or median (such as in steps 5, 9, and 11).

Q1 The mean and median are both 3.5.

Q2 Because each outcome is equally likely, you would expect to roll a 1 about 100 times of the 600. The mean of the face values should be close to 3.5, and the median will likely be 3 or 4.

INVESTIGATE

6. Holding down the Shift key forces Fathom to treat the data as categorical as opposed to numeric, so now 1 is treated as a discrete category rather than as a number. Notice that there is also a way to force Fathom to treat a non-numeric attribute as numeric.
7. Being able to model rerolling the dice helps demonstrate the variation experienced. The faces do occur about 100 times each, but the graph is probably more ragged than some students might expect.

Q3–Q4 In general, the graph has six bars, each with height close to 100 (plus or minus 25 or so). The mean is close to 3.5 (plus or minus 0.2 or so). The median is 3 or 4 with equal likelihood, but every now and again, when the number of rolls from 1 to 3 is exactly equal to the number of rolls from 4 to 6, the median is 3.5 (perhaps 3 times out of 100 tries).

Q5 Mean and median will be close to 7.

Q6 The median is almost always 7 for a large number of rolls. The mean is close to 7 (plus or minus 0.3 or so). Actual observations will vary.

Q7 The range is $12 - 2$, or 10. It does not change.

Q8 The mean is close to 7 (plus or minus 0.2 or so).

Q9 Students will probably know that 7 is the most likely outcome for the sum of two dice and that 2 and 12 are the least likely, and that choosing a random number between 2 and 12 is not good for simulating the sum.

Q10 randomInteger (1, 6) + randomInteger (1, 6)

Q11 The mean is 7, the median is 7, the minimum value is 2, the maximum value is 12, and the range is $12 - 2$, or 10. These summary statistics are the same values as for *Roll_2_to_12*.

In your discussion of this answer, you might ask about the need for other summary statistics that will be different for these two different data sets.

Q12 The graph of *Roll_2_to_12* has 11 bars of roughly equal height, whereas on the graph of *Add_Rolls*, the middle bar (at 7) is the tallest, and the bars on either side get progressively shorter until 2 and 12, which have a height usually somewhat less than 30 for 1000 rolls. The graph of *Add_Rolls* is more bunched up around the center; thus, the graph of *Roll_2_to_12* has a more even distribution.

Q13 It's a good model.

EXPLORE MORE

1. This distribution would have a mean and median of 7.5 and would be even more mounded or bunched up in the center. Although the distribution of random integers from 2 to 12 is basically flat and the distribution of the sum of two 6-sided dice is like a tent (two lines meeting), this distribution begins to show the bell shape.
2. The standard deviations observed for *Roll_2_to_12* will likely be between 3.05 and 3.25, whereas the standard deviations for *Add_Rolls* will likely be between 2.3 and 2.6. This reflects the fact that the graph of *Roll_2_to_12* is significantly more spread out than the graph of *Add_Rolls*.

EXTENSIONS

1. Write a formula for the mean of the face value of a die that is numbered consecutively from a to b and rolled a large number of times (n). Check your formula against your answer to Q2 (when $a = 1$ and $b = 6$) and your answer to Q5 (when $a = 2$ and $b = 12$). Your formula will be in terms of a and b.

   Answer: mean $= \frac{a+b}{n}$
2. Design a single die that would generate the same kind of distribution as the sum of two 6-sided dice. How many faces would it need? Which face values would be used, and how many times would each be repeated?

   Sample answer: It would have 36 sides with 2 and 12 each appearing once; 3 and 11, twice; 4 and 10, three times; 5 and 9, four times; 6 and 8, five times each; and 7, six times.

Box Plots—Insurance

In this activity, you will advise an international insurance company how much to charge for life insurance in various countries. A company selling life insurance must be sure to earn more money in premiums than it pays in death benefits. To accomplish that goal, the company must predict how long customers will live. It can't know when any particular customer will die, but it can gather data about the average (mean) lengths of life for persons in various groups. These data are called *life expectancies.* Groups of people who have a longer life expectancy are charged lower life insurance premiums.

Q1 Who do you think lives longer on average, men or women? Europeans or Latin Americans? Why?

INVESTIGATE

1. Open the sample document **Insurance.ftm.** Scroll through the table to find the total number of countries recorded.

Graph

2. Drag a graph from the shelf and make a dot plot of all the life expectancies, either for men or for women.

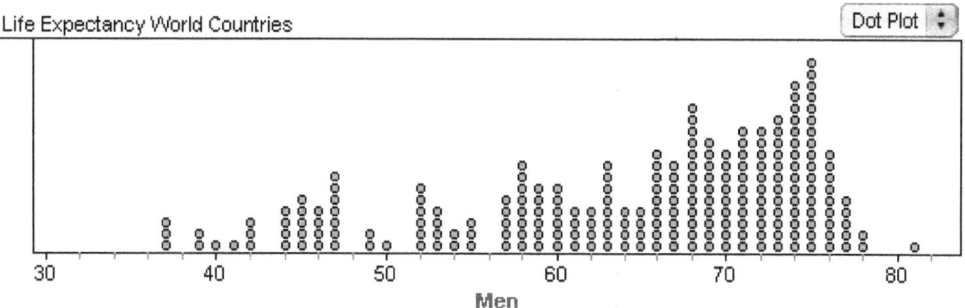
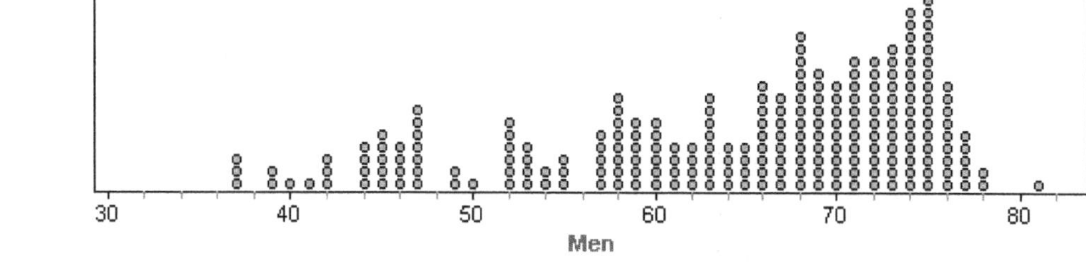

To find the dot that represents the United States, highlight United States in the table. The corresponding dot will be highlighted.

United States (73)

3. Move your mouse over the dots in the dot plot and notice that the country and life expectancy are displayed in the status bar at the bottom left of the window.

Q2 Determine the maximum, minimum, and range of the life expectancies. Which country has the maximum life expectancy? The minimum?

Q3 What is the value in the middle of the range? Do you think the mean and median will be close to the middle of the data? Why or why not?

Q4 Do you expect the mean to be greater or less than the median? From the graph, what values do you predict for the mean and median? Explain your reasoning.

Data like these are called *skewed left* because instead of being symmetric, the graph has a tail of data to the left.

Q5 In which countries would you advise that life insurance premiums be the lowest? The highest?

4. Because the data are broken down by sex, you suspect that recommendations for the sexes might differ. Drag a new graph from the shelf and create a dot plot of life expectancies for the other sex.

Q6 Compare the dot plots for men and women. Identify at least two similarities and two differences between the dot plots.

5. A different kind of graph, called a *box plot,* allows other comparisons. To change a dot plot to a box plot, click on **Dot Plot** in the top right corner of the graph window and choose **Box Plot.** Box plots, sometimes called *box-and-whisker plots,* are based on the five-number summary—minimum, first quartile, median, third quartile, and maximum.

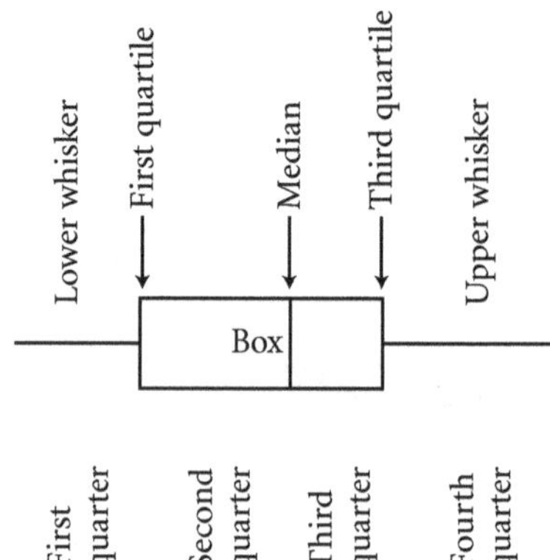

If you move your mouse over these quarters, Fathom displays information about them in the status bar (for example, "Lower whisker from 37 to 58"). The lower whisker is the first quarter, and the upper whisker is the fourth quarter. When you click on a quarter, the countries represented will be selected in the table and in both graphs. Click on each of the four quarters in one box plot and notice which countries are highlighted on the other box plot.

Q7 Compare the four quarters on the two box plots. Is there evidence that one sex tends to live longer than the other? If so, does that tend to be true within the individual countries as well as overall?

6. Choose **Graph | Plot Value** to plot the mean and the median on each box plot.

Q8 How good was your prediction in Q3? How do the mean and median compare for data that are skewed left?

7. Comparing two box plots can be difficult because the scales are different. To make this easier, you can put the two plots on the same graph. Drag the *Men* attribute from the table to the box plot for *Women,* aiming for the horizontal axis. A box will appear with a plus sign to the left. Drop the attribute on the plus sign. The mean for *Men* will appear on the graph.

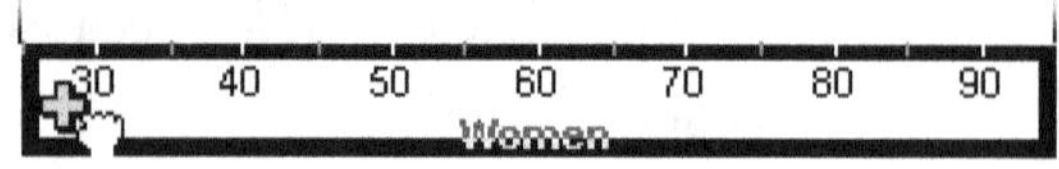

Q9 What do you notice that you couldn't see before?

8. Choose one region of the world. Scroll through the table to find that region and the total number of countries recorded.

Q10 Which region did you choose?

9. To make a graph of data from just the region you chose, select the box plot with life expectancies for both men and women and choose **Object | Duplicate Graph.** Drag the top border of the new graph so that you can easily compare the two graphs. Then choose **Object | Add Filter.** Enter a formula that limits the data to your region; for example, if your region is Latin America, enter the formula Region = "Latin America" and click **OK.** Your new graph will contain data only for the countries in your region.

Q11 How does the box plot of your region compare with the box plot of world data? Explain any differences. In particular, consider means, medians, and skew.

Q12 Should life insurance premiums for your region be more or less than those for the world as a whole? Defend your recommendation.

10. Examine at least one other region of the world. First, duplicate the graph for your region. Then, double-click the old filter and change the formula to limit data to the new region.

Q13 Compare the data for this region with the data from your original region. Focus on mean, median, and skew. What advice would you give about life insurance premiums in the new region? Defend your opinions by referring to the graph.

EXPLORE MORE

1. Investigate how life insurance premiums in the other regions should compare with your original region.

2. Create a new attribute called *Difference* with the formula Women – Men. Create the dot plot for this new attribute. Which country has the largest difference? Which has the smallest difference? How many countries have a negative difference? What does a negative difference mean? Is difference skewed left, skewed right, or symmetrical?

Objective: Students will use Fathom's ability to quickly graph data, then they will compare box plots and characterize the data as skewed left, skewed right, or symmetrical. Fathom's filters allow students to consider a subset of the data.

Student Audience: Pre-algebra, Algebra 1

Activity Time: 35–50 minutes

Setting: Paired/Individual Activity or Whole-Class Presentation (use **Insurance.ftm** for either setting)

Mathematics Prerequisites: Students can calculate the range of a data set.

Fathom Prerequisites: Students can create new graphs and plot values on graphs.

Fathom Skills: Students will change the graph type, create a box plot with two attributes plotted on the same scale, and add a filter.

Notes: Help students make connections as they see different ways of graphing and comparing data. Dot plots show the shape and spread, including skewness. Box plots draw attention to the middle of the data and show the spread but not the shape. How does each graph affect the comparisons of groups?

Ask groups who finish early to think about and present their ideas about the ability of the people from countries in their answer to Q5 to buy insurance. Would the company be likely to sell insurance in countries where the premiums would be highest?

As you listen to students talk about quarters, be aware that some will use the word *larger* to mean "more spread out." Use the more exact wording, *more spread out,* and discuss how the box plot's area seems to convey more than it actually represents.

Check that students are aware that axes need to be the same when comparing graphs.

This is an excellent opportunity for several groups to share answers to the same questions. Take time for these group presentations to increase your students' ability to communicate about the relationships, the mathematics. Their presentations can lead to interesting discussions about what can be seen in different kinds of comparisons.

For a Presentation: As you present the activity, ask the student running the computer to create dot plots for both men and women, then skip step 4. Once students have given their opinion on the relationship between the mean and the median in Q4, you might plot the values of the mean and median on each dot plot. As students answer Q6, encourage many different answers. At Q10, to maximize the number of comparisons students will find as they answer Q13, ask for two or three regions and complete step 9 for those regions.

Q1 Answers will vary. In their predictions, students might consider risky activities, health, standard of living, and other factors.

INVESTIGATE

1. There are 229 countries recorded in the data set.

Q2 For men: The maximum life expectancy is 81 years in Andorra; the minimum is 37 years in three African countries; and the range is $81 - 37$, or 44 years.

For women: The maximum is 87 years in Andorra; the minimum is 37 years in Zambia and Malawi; and the range is $87 - 37$, or 50 years.

Q3 For men: The value in the middle of the range is $37 + \frac{44}{2}$, or 59 years. The mean and the median will be more than 59 years, because most of the data points are more than 59 years.

For women: The value in the middle of the range is $37 + \frac{50}{2}$, or 62 years. The mean and the median will be more than 62 years, because most of the data points are more than 62 years.

Q4 The mean will be less than the median, because there are a few values much less than most of values, but there are no corresponding values much greater than the bulk of values.

Q5 Andorra has the highest life expectancy, so the insurance premiums should be lowest; Malawi and Zambia are among the countries with the shortest life expectancy, so their premiums should be highest.

Q6 The two dot plots are both skewed left, and the minimum values are the same (37). The dot plot for *Women* has many more dots above 80, its maximum value is larger, and it has fewer dots in the 60s and more in the 70s. (Other comparisons can be made.)

5. If students are having difficulty with the concept of a box plot, create a dot plot and a box plot of the same data. Click on each quarter on the box plot to highlight dots in red and then count the number of corresponding dots in the dot plot (in the diagram, the third quarter was selected on the box plot and the corresponding red dots appear without borders in art below). Each time, one-quarter of the dots will be highlighted on the dot plot, and the range of the highlighted dots will be determined by two numbers from the five-number summary for the data.

   (*Note:* Some textbooks define and calculate the quartiles—the endpoints of quarters—in slightly different ways.)

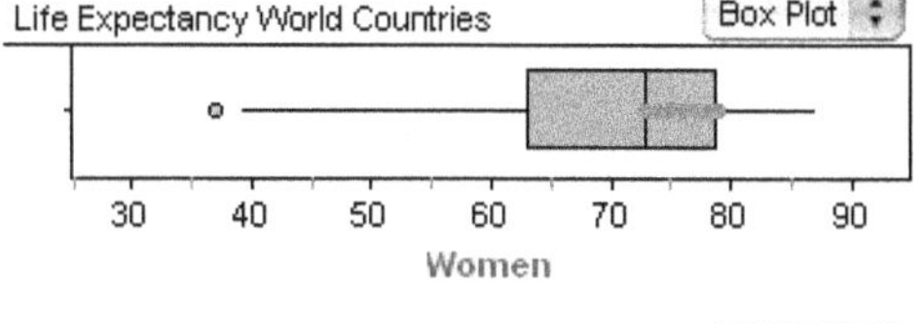

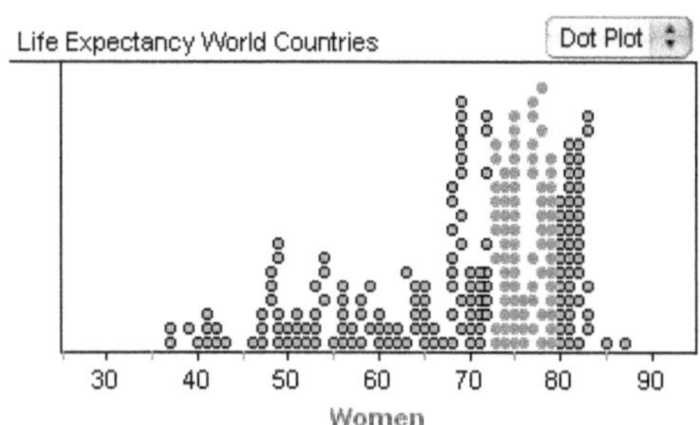

Q7 The five-number summary for *Men* is 37, 58, 68, 73, and 81, and for *Women*, it is 37, 63, 73, 79, and 87. The median life expectancy for women is five years more than for men. Although the minimum values for the two are identical, the rest of the numbers for women are five or six years greater than the corresponding number for men. There is some evidence that, on average, women live longer than men. Most of the countries in a particular quartile of one box plot are in the corresponding quartile in the other box plot. This indicates that the life expectancy for men and women are ordered similarly.

Q8 The mean is smaller than the median. The mean for men is 64.2 years, and the median is 68 years. Comparisons with students' predicted values will vary. Data sets that are skewed left have means that are less than their medians.

7. When the two attributes are graphed on the same box plot, the results will be similar to this one.

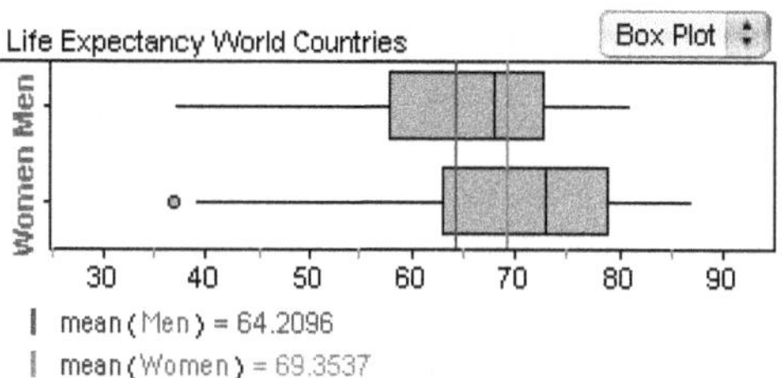

Q9 Though the women's data are more spread out, the middle half of the data is equally spread out for both groups. The third quartile for *Men* is equal to the median for *Women.*

9. The duplicated graph will be similar to this one.

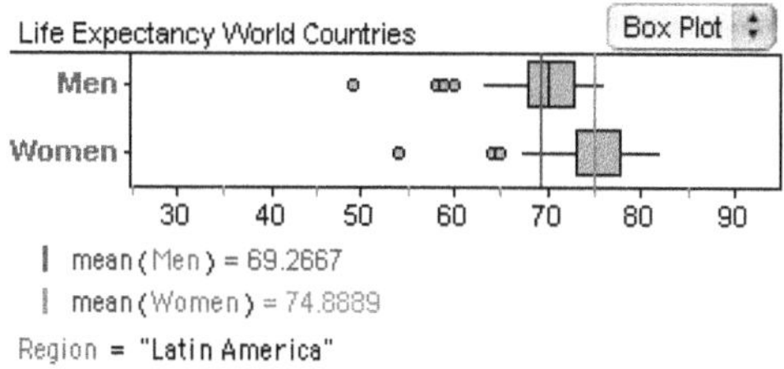

Q10–Q12 Answers will vary.

Q13 Sample answer: The middle half of the data for Africa is more spread out than it is for Latin America. Both the means and the medians are higher in Latin America. In Africa, most of the countries have short life expectancies, but a few have much larger values (skewed right); in Latin America, however, most of the countries have long life expectancies, with a few having smaller values (skewed left).

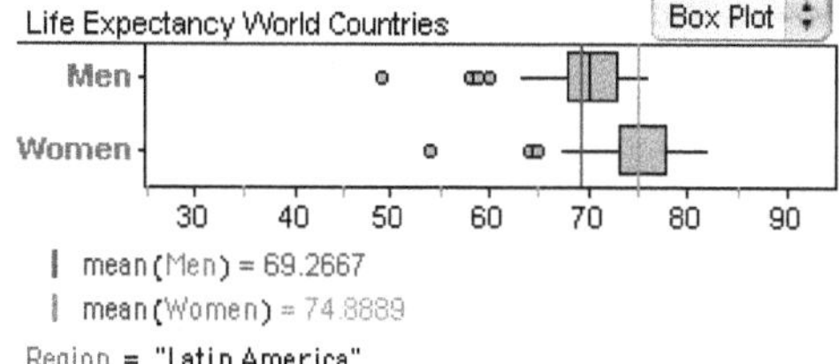

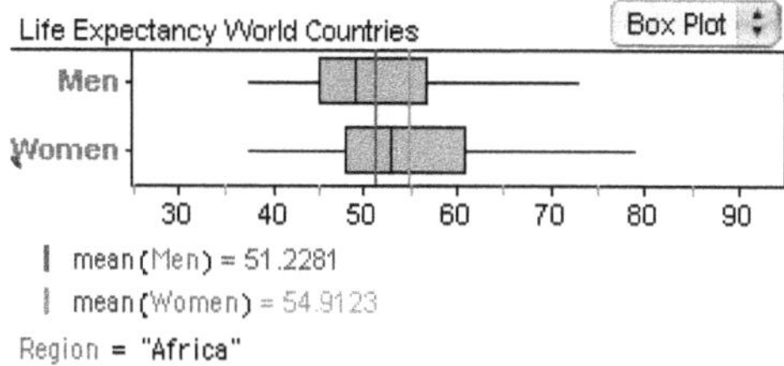

EXPLORE MORE

1. Many different graphs and calculations can be made.

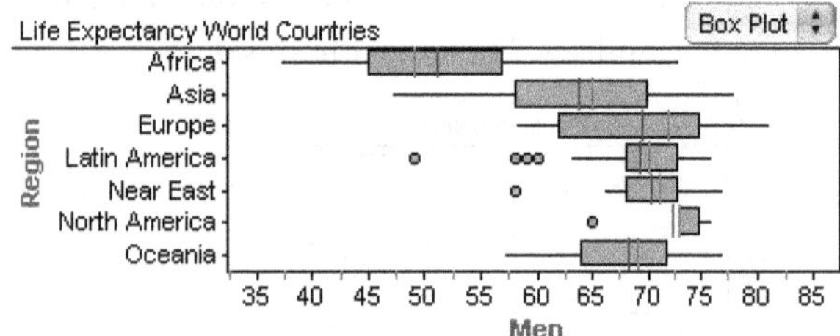

Although the median life expectancy in Europe is higher than in Latin America, the means are comparable; otherwise, the most striking difference is the greater spread of the European values.

2. Largest are Belarus, Latvia, and Russia. The attribute *Difference* is smallest for Vanuatu. Nepal, Bangladesh, Zimbabwe, Zambia, and Malawi have the smallest difference in life expectancies. Five countries have a negative difference, meaning the life expectancy for men is higher than for women in those countries. The distribution is approximately symmetric.

EXTENSION

Most people can't afford to purchase life insurance. Find data on life expectancies for people of higher incomes in the various countries and regions and adjust the recommendations you'll make to the insurance company.

Box Plots—Think Box

It can be fun to see how far you can go without anyone noticing. In this exploration, you'll play some games to see how much you can push around data points without changing their box plot.

Q1 Draw a box plot on your paper for the data set {0, 1, 2, 3, 4, 5, 6, 7, 8, 9, 10, 11, 12}. How much would the box plot change if the 12 became a 13?

INVESTIGATE

1. Open the document **Think Box.ftm.** You will see a collection, Original Data, together with a dot plot and a box plot of the data.

Neither method is wrong; they just follow slightly different rules.

Q2 Is the box plot that Fathom created the same as the one you completed by hand in Q1? It might not be. Some books outline methods of graphing box plots that are different from those used by Fathom. If your plot is different, explain how Fathom calculates the five numbers determining the box plot.

2. You will also see an exact copy of the original data in the Smallest Mean collection, together with its dot plot and box plot for the first game. Notice that the mean is computed.

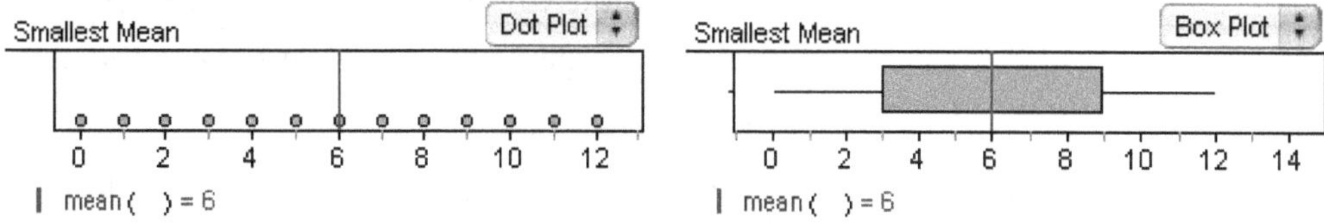

Q3 What is the mean of the data?

Here are two games to play with the data in the Smallest Mean collection. The object of both games is to move dots in the dot plot to get the smallest and largest values of the mean. Here are the rules.

The Rules of Game 1

- You can drag only one dot at a time in the dot plot.
- You cannot drag a dot past another one in the dot plot.
- You cannot change any of the five-number summary values (minimum, first quartile, median, third quartile, or maximum). That is, the box plot should always look exactly like the box plot for the original data. The mean is not part of the five-number summary and is free to change.

3. Play Game 1 to find the smallest mean.

Q4 What strategy did you use in playing the game?

Q5 What is the smallest value of the mean? Check your answer against those of others in the class. What numbers give the smallest mean?

Q6 As another check, calculate the mean of your Smallest Mean data by hand and write it as a fraction in lowest terms.

$$\frac{0 + 0 + 0 + 3 + 3 + \cdots + 9 + 9 + 12}{13}$$

Highlight the original collection and choose **Object | Duplicate Collection.**

4. To find the largest value of the mean, make another copy of the data collection. Double-click the name of this new collection and call it Largest Mean.

Graph

5. Drag a new graph from the shelf. Double-click the Largest Mean collection to show its inspector. Drag the attribute *Value* to the graph to create a dot plot.

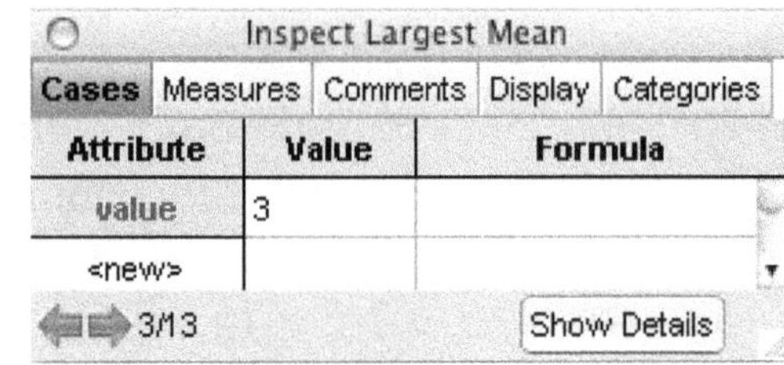

Use **Graph | Plot Value** to plot mean( ). Now drag down a new graph and make the corresponding box plot. You have duplicated the original situation.

Q7 Find the largest possible value of the mean, following the rules of Game 1. What values will make the largest possible mean? Write the mean of these values as a fraction in lowest terms.

Q8 What is the difference between the largest and smallest possible values of the mean for Game 1? Express this as a fraction in lowest terms. Devise a shortcut for finding this difference.

Q9 Compare the completed dot plots for Smallest Mean and Largest Mean. Which of the two graphs has the bigger spread? Justify your answer.

The object of Game 2, like Game 1, is to find the largest and smallest values of the mean, but the rules are a little different.

The Rules of Game 2

Follow all the rules of Game 1, except you may change the value of the median. None of the other four values of the five-number summary can change.

6. Click on the dot plot for the Largest Mean collection and use **Graph | Plot Value** to plot median( ).

Q10 Use this dot plot to play Game 2, finding the largest possible value for the mean under the relaxed rules. What strategy did you use? What is the largest possible value for the mean now? What is the median value when the mean is the largest possible? Is this graph skewed left or skewed right?

Q11 Without moving any dots, determine the smallest possible mean and the corresponding value of the median.

EXPLORE MORE

1. Find examples of data sets that can be reconstructed from seeing just the box plot and knowing how many data points are in the data set. Make some conjectures about the kinds of sets that can be reconstructed from their box plots. Find examples of data sets that can be reconstructed from knowing the number of data points, the box plot, and the mean.
2. Create a data set and a set of rules different from those in this activity, then challenge other students to play.

Box Plots—Think Box

Objective: Using the ability to drag data points in Fathom, students will explore summary data and realize that it does not reflect all the information in a data set. Because Fathom allows students to change data by dragging dots on a dot plot, students can visualize changes that affect the box plot.

Student Audience: Pre-algebra, Algebra 1

Activity Time: 40–60 minutes

Setting: Paired/Individual Activity (use **Think Box.ftm**) or Whole-Class Presentation (use **Think Box Present.ftm**)

Mathematics Prerequisites: Students can calculate the five-number summary for a data set and create a box plot.

Fathom Prerequisites: Students can create new dot plots and box plots, duplicate objects, and plot the value of statistical measures on graphs.

Fathom Skills: Students will rename a collection and use the inspector as a source of data attributes.

Notes: Box plots are useful for describing the center and spread of data, but the data can change without affecting the box plot. To set values precisely, students might find that double-clicking on a dot and typing the exact value in the inspector works best.

As you watch pairs of students work through the activity, choose groups that have slightly different answers to present their methods and results. Move down to the lowest mean and up to the highest in the order of presentations. Talking about methods is as important as talking about results.

For a Presentation: To orchestrate a whole-class discussion, use **Think Box Present.ftm,** which already has the Largest Mean collection. Take suggestions from the class on which dots the student running the computer should move. Build your discussion around the questions, asking students to make predictions and then using Fathom to check them out. You might assign Explore More 1 as homework after the whole-class discussion.

Q1 Student box plots may look like the one in **Think Box.ftm.** Fathom calculates the five-number summary as 0, 3, 6, 9, and 12; the median is considered part of both the lower half of the data and the upper half of the data. If 12 became 13, the right whisker would become 1 unit longer.

INVESTIGATE

Q2 Answers will vary. For example, Fathom includes the middle value, 6, in the lower half of the data when calculating the first quartile, but some conventions identify the first quartile by finding the middle number of those numbers less than 6.

Q3 The mean of the data is 6.

Q4 As soon as 0, 3, 6, 9, or 12 is moved, the box plot changes shape and the rules are broken. Move the two dots just above 0 down to 0, those just above 3 down to 3, and so on.

Q5 The smallest mean is about 5.07692.

Q6 Using the numbers 0, 0, 0, 3, 3, 3, 6, 6, 6, 9, 9, 9, 13, the smallest value for the mean in lowest terms is $\frac{0+0+0+3+3+3+6+6+6+9+9+9+12}{13} = \frac{66}{13}$.

Q7 Using the values $0 + 3(3) + 3(6) + 3(9) + 3(12)$, the largest possible value for the mean is $\frac{0 + 3(3) + 3(6) + 3(9) + 3(12)}{13} = \frac{90}{13}$.

Q8 The difference between the largest and smallest possible values is $\frac{90}{13} - \frac{66}{13} = \frac{24}{13}$. Both means have three each of 3, 6, and 9, so the difference will depend only on the 12; that is, $3(12) - 12$, or 24, will be in the numerator and 13 will be in the denominator.

Q9 The two dot plots are mirror images of one another with the value 6 as the axis of reflection. Their spreads will be identical.

Q10 The largest possible value for the mean is $\frac{99}{13} \approx 7.61538$, and it occurs when the median is 9. Move the data points that are not 0, 3, 9, or 12 as far to the right as possible without moving any dot past its neighbor. Six dots will be at 9, making the median 9.

Q11 Starting with the Smallest Mean collection, move the three dots at 6 to the left by 3 units. This will make the total go down by 9, from 66 to 57. The smallest mean for Game 2 is $\frac{55}{13} \approx 4.38462$ and will occur when the median is 3.

EXPLORE MORE

1. Data sets whose cases can be determined from a box plot include those with just five cases, those with six or seven cases and the median equal to one of the quartiles or with outliers, and those with any number of cases and all five-summary values the same. If you also know the mean, you can identify any six-member data set from its five-number summary.
2. Answers will vary.

EXTENSIONS

1. Compare the spread of the two graphs created for Game 1 and the graph created for Game 2. Plot the standard deviation, **s( )**, on the two graphs. The graph with the greater spread should have the greater standard deviation.

   Answers:

   Game 1 smallest mean: s ≈ 3.95

   largest mean: s ≈ 3.92

   Game 2 s ≈ 4.0

2. Create a measure for spread that produces the same result for the data {0, 0, 0, 3, 3, 3, 6, 6, 6, 9, 9, 9, 12} as it does for {0, 3, 3, 3, 6, 6, 6, 9, 9, 9, 12, 12, 12}.

   Answers will vary. Sample answer: Measure the spread by taking the sum of the distances from the median and divide by the number of elements:

   $\frac{1}{n}\sum_{i=1}^{n} |\text{median} - x_i|$, or $\frac{1}{13}(6 + 3 + 3 + 3 + 0 + 0 + 0 + 3 + 3 + 3 + 6 + 6 + 6) \approx 3.23$

Histograms—Ontario Communities

Governments collect census data to answer interesting questions about their populations. For example, Statistics Canada reports the median 2001 income in each community in the province of Ontario. What do the data say about communities in which the average annual income is above the median?

Q1 How much money do you think the average person (15 years old or older) in your own community makes in a year?

Q2 Do you think annual income is higher in cities than it is in towns? Explain.

INVESTIGATE

1. Start Fathom and open **Ontario.ftm.**

The scatter plot shows 445 communities in Ontario, colored by median income. For example, in each community colored red, the median income is greater than about $30,000.

Q3 How does the scatter plot relate to the map of Ontario?

Q4 On the scatter plot, about how many of the 445 communities have incomes greater than $30,000?

Double-click the graph to show the inspector, or add a case table for the Census Subdivisions data and drag the attribute from the table.

2. It's difficult to count the communities. Create a new dot plot of the attribute *Median_Income_15plus.* You can drag the attribute name from the graph's inspector.

Q5 Is the dot plot skewed left, symmetrical, or skewed right? What does the shape indicate about whether the mean or the median will be larger?

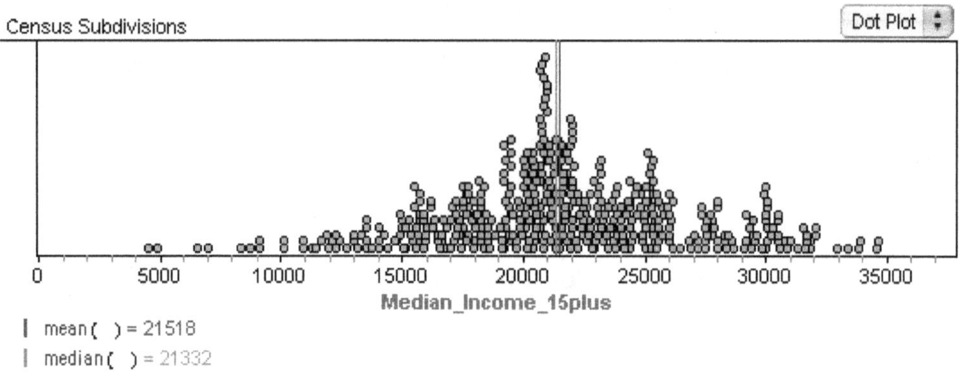

With the graph selected, choose **Graph | Plot Value** and type mean( ); repeat for median( ).

3. Plot the value of the mean and the median on the dot plot.

Q6 Are the values related in the way you indicated in Q5? You may need to drag to adjust the horizontal scale until you can see both lines.

Q7 Is the median you plotted the median income for the entire province? Explain.

Q8 From the dot plot, try to determine the number of communities that appear to have a median income greater than $30,000.

4. It may be easier to count these communities in a histogram. Choose **Histogram** from the pop-up menu at the top right of the graph.

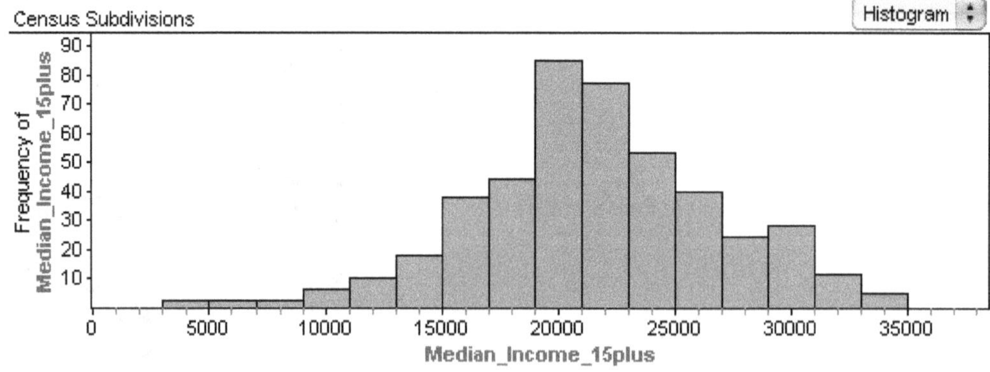

The general shape of the histogram is similar to the shape of the dot plot, but now the data have been collected into groups, often called *bins.* The frequency, or number of communities in each bin, is represented by the height of the bin. You can see this frequency by looking at the scale on the vertical axis or by holding the cursor over the bin and looking at the status bar. You can click on the bin to see those communities highlighted on the scatter plot.

Q9 Move your mouse over the tallest bin and observe the information displayed at the bottom left of the Fathom window. In what interval of incomes does the greatest number of communities lie? How many communities have incomes in this interval?

5. Move your mouse between two bins until a double-tipped horizontal arrow is displayed.

   Drag to change the bin widths. Notice that the vertical scale does not change as you do this. To get a better vertical scale, you can drag the scale details or simply reselect **Histogram** as the graph type.

Q10 Adjust the histogram so that exactly eight bins are displayed and you can see all the data. Now what interval of incomes has the most communities in it? Do the mean and median values lie within this interval? How many communities are in this interval? From this graph, can you determine the exact number of communities with median incomes greater than $30,000?

6. You may need more precise control over the histogram's display. Double-click on any white space on your histogram to show the graph inspector (or choose **Object | Inspect Graph**). Adjust the *binWidth* and *binAlignmentPosition* fields until the last bin starts at $30,000. You may need to experiment with a few values to figure out how these fields affect the histogram. The graph will adjust after you change a field's value and press **Enter.** There is no need to hide the inspector. Select the bin or bins representing median incomes of $30,000 or more, and they will turn red.

Q11 What values of *binWidth* and *binAlignmentPosition* are you using? How does the exact count compare with your earlier estimates? What percentage of communities have a median income of $30,000 or more?

7. To discuss locations of the communities with the highest incomes, look at the original scatter plot and notice the communities that are selected.

Q12 Where on the map are the communities that have the highest median incomes? Where are the communities with the lowest median incomes?

EXPLORE MORE

1. Duplicate your histogram and add a filter Population_2001 > 50000. How many communities have a population greater than 50,000? What percentage of the communities with population greater than 50,000 have a median income of

$30,000 or greater? How does this compare with your percentage in Q11? Does this provide evidence that people in cities make more money on average? Explain.

2. Go back to the scatter plot of Ontario that shows the latitude and the longitude of each community. Drag an attribute other than *Median_Income_15plus* to the center of the graph. What does this tell you about the population of Ontario? For example, how is family income or the value of dwellings distributed across the province?

Histograms—Ontario Communities

Objective: Students will interpret data displayed in scatter plots and histograms. In Fathom, students explore how changing data affects the five-number summary used to create a box plot and how a histogram changes as the bin width changes.

Student Audience: Pre-algebra, Algebra 1

Activity Time: 20–35 minutes

Setting: Paired/Individual Activity (use **Ontario.ftm**)

Mathematics Prerequisites: Students can characterize a data set as skewed left, skewed right, or symmetrical.

Fathom Prerequisites: Students can open a Fathom document, use inspectors, create dot plots, and plot values on graphs.

Fathom Skills: Students will create histograms, adjust the display of a histogram, and select the cases within a bin of a histogram.

Notes: As you talk with students, ask them what they see in each type of display. What does the display show best? Which display is best to answer a specific question? Which questions does a particular display answer? As students give answers to Q7, ask why. For questions whose answers vary, choose a variety of displays to be shared. As you ask students to share a particular histogram, make sure they are able to copy that display and put it aside where it won't get changed. As students share, you might bring up the question of which histogram fairly portrays the data and for what purpose. Extensive information on the sources of the data in this activity is given in the Census Subdivision inspector **Comments** panel. A central point in the community was used for the latitude and longitude measurements.

Q1 Answers will vary. At the time of the census, the Canadian dollar was worth less than $0.75 American dollars, so CAN$30,000 would be around US$22,000.

Q2 Answers will vary. Often, the suburbs have the highest income, and the answer might depend on how suburbs are treated as communities.

INVESTIGATE

Q3 Communities are concentrated in the large urban centers around Toronto, Ottawa, and Windsor, as well as along a line of industrial towns in northwest Ontario.

Q4 Answers will vary.

Q5 The dot plot appears approximately symmetrical. The mean and the median will be relatively close to one another.

Q6 The mean and median are within $200 of each other, with the mean larger than the median.

Q7 No; this is the median of the median values in the communities. Because all communities are not the same size, large communities are underrepresented in the median of the medians.

Q8 It is difficult to count these dots, even if students resize the graph. There appear to be approximately 30 communities. By creating a table and sorting by that attribute in the table, students will see that 29 communities have median incomes of greater than $30,000.

For example, the greatest number of communities in the histogram lies in the range 19,000 ≤ *Median_Income_15plus* < 21,000, which contains 85 communities.

Q9 Answers will vary. One possible answer is that the greatest number of communities lies in the range 19,000 ≤ *Median_Income_15plus* < 21,000, which contains 85 communities, or 20,500 < *Median_Income_15plus* ≤ 25,000, which contains 163 communities. Students would be very lucky to create a histogram with $30,000 as one of the divisions.

Q10 Answers will vary. Most students will need to adjust the histogram further.

Q11 One possible set of values is *binWidth* = 5000 and *binAlignmentPosition* = 0. There are 29 communities with median income of $30,000 or more. The percentage of communities in this range is $\frac{29}{445}$, or approximately 7%.

Q12 Most of the higher incomes are found around the Great Lakes. A few of the communities with a low median income are in urban areas, but most of them are more remote.

EXPLORE MORE

1. Thirty-seven communities have a population greater than 50,000. Eight of these communities have median incomes that are $30,000 or more. Thus, $\frac{8}{37}$, or approximately 22%, of the larger communities have larger incomes, compared with only 8% of all communities. This means that if you are in a larger community, you are about three times more likely to be in a community with a median income of $30,000 or more.
2. Answers will vary.

EXTENSION

Use other attributes in the collection to compare the median incomes of males 15 and older with females 15 and older. Is there evidence of a difference? If so, what factors might contribute to this difference?

Ratios—Surveys

Your teacher will ask your class for opinions on some questions. Your goal is to use the responses of your classmates to estimate how all the kids in your school would answer those questions.

Q1 Before you see the actual results, guess how many students in your class answered *Yes* to the first question and how many answered *Yes* to the second question. Now write at least three statements that summarize your guesses in different ways.

One way to compare two things is to use a ratio.

Q2 For the first question your teacher asked, give two different ratios that include your guesses as to the number of *Yes* votes. Do the same for the second question.

INVESTIGATE

You can set up a data collection by entering data into a Fathom case table.

1. Open Fathom and drag a new case table from the shelf into the Fathom window.

You might choose different attribute names or omit the underscore between the *Response* and the *1* or *2*, but attribute names can't contain spaces.

2. To set up an attribute, click on <*new*> in the table and enter Response_1 for the attribute. Enter another attribute, Response_2. If your teacher asked three questions, you would add a third attribute, Response_3. Your teacher will now provide you with the results of the survey. Enter Yes or No (or Y or N) for each student response to the first two questions. When you are done, check that you have entered the correct number of responses.

3. Graphs will help you visualize and interpret the data. Drag a new graph from the shelf. Drag the attribute *Response_1* from the case table to the graph. Drop the attribute under the graph's horizontal axis. You may be surprised to see a bar chart rather than a dot plot.

You get a bar chart because the responses are not numbers, they're *categories.* A bar chart allows you to see the number of data points in each category. By comparing the lengths of the bars, you can see which category holds more responses.

32 cases (80.0%) are No.

4. To see the actual numbers, watch the status bar at the bottom left of the Fathom window while you move the cursor over the bar that represents the *No* responses and then on the bar that represents the *Yes* responses.

Q3 How many students answered *No?* How many said *Yes?*

The graph must be selected for the option **Object | Duplicate** to appear.

5. To approximate how much each category contributes to the whole, you might make a ribbon chart. Make a copy of the bar chart by choosing **Object | Duplicate Graph.**

Move the new graph to the side. Use the pop-up menu in the upper right corner to change the graph from a bar chart to a ribbon chart.

Q4 What fraction of the ribbon represents *Yes* responses?

Q5 Look again at the bar chart. Estimate a fraction that describes how large the *Yes* bin is compared with the *No* bin.

Q6 Calculate each of the following ratios. Express your answer as both a reduced fraction and a decimal.

a. $\frac{\text{Number of } Yes}{\text{Total number}}$ b. $\frac{\text{Number of } No}{\text{Total number}}$ c. $\frac{\text{Number of } Yes}{\text{Number of } No}$

d. $\frac{\text{Number of } No}{\text{Number of } Yes}$ e. $\frac{\text{Total number}}{\text{Number of } No}$ f. $\frac{\text{Total number}}{\text{Number of } Yes}$

Q7 Use complete sentences to explain the two ratios that compare the number of *Yes* responses (in the numerator) to another quantity.

Q8 Which chart was better in helping you visualize each ratio?

6. Analyze the results for the second question. Make a bar chart and a ribbon chart with *Response_2* on the horizontal axes.

Q9 Use at least two ratios to summarize the class responses to the second question.

To use the data from your class to estimate how many students in your school would answer *Yes* to these questions, you'll be assuming that the responses from the whole school would have the same ratios as the responses in your class.

Q10 Estimate how many *Yes* responses to each question would come from your entire school. Be sure to justify your reasoning and be clear about which ratio you chose and how you used it.

Q11 Estimate how many *No* responses to each question would come from your entire grade. Explain.

EXPLORE MORE

1. If your teacher asks your class a question that has three answers, use graphs and ratios to estimate how many students in your school would give each answer, assuming that your class is representative of the school.
2. Look for patterns in the responses. For example, did students who answered *Yes* on the first question also answer *Yes* on the second question?
3. Your own class might not be very representative of the entire school. How might you select a small group that would be more representative of all the students?

Objective: Students deepen their understanding of ratios as they look at different ratios from the same information. They will look at different ways to represent those ratios in bar charts and ribbon charts, which are quickly created in Fathom, and will use different ratios to solve proportions for an unknown.

Student Audience: Pre-algebra, Algebra 1

Activity Time: 25–40 minutes

Setting: Paired/Individual Activity (based on survey results from the whole class)

Mathematics Prerequisites: Students can reduce a fraction and convert a fraction to a decimal.

Fathom Prerequisites: Students can duplicate Fathom objects.

Fathom Skills: Students will open a case table and create a new attribute, a bar chart, and a ribbon chart.

Notes: Start the activity by asking two yes-no questions, such as those listed below. Students might write their answers on slips of paper. If you want students to do Explore More 1, ask a third question with three answers. Any questions will work, as long as not all students will give the same answer. While you gather and tally the responses, students can work on Q1 and Q2.

Report the responses student by student, providing the data for the table case by case. The quickest way might be to ask two students to work quickly to enter the data into a Fathom document, then have the whole class access that document. Alternatively for two questions, you could tally the ballots for the number of YY, YN, NY, and NN so students can quickly enter data and copy responses down rows.

You want questions for which students would like to hear the opinion of their peers, yet that don't reveal personal or sensitive information. Here are some suggestions:

- Should there be a student uniform that all students are required to wear?
- Would it be better to have fewer classes with longer periods? (Or the opposite question if you are on a block system.)
- Should the cafeteria serve more healthful options and fewer junk foods?

The third question for Explore More 1 should have three responses, such as *Earlier, Later,* or *Same time.* Here are some examples:

- Should school start and end earlier, later, or at the same time?
- Should (insert school function: dance, football game, band concert) start earlier, later, or at the same time?
- If you were in charge of the TV network, would you move the evening news (or a current show or a student-designated favorite show) to start earlier, later, or at the same time?

Many ratios can come from the same information. Using the right ratio at the right time is an important skill. If students simply take any ratio given and use it, they will often have poor results.

Different graphs of the same information can help students visualize that information in different ways. Both bar charts and ribbon charts can be used to display categorical information, but each allows you to see different ratios. Give students opportunities to explain why they are using a certain ratio or proportion. As you circulate, ask groups what they have done and why. During presentations, student groups can show their charts and give opinions on Q8.

Q1 Answers will vary.

Q2 Students could give the ratio of *Yes* to total responses, the ratio *Yes* to *No,* the ratio of total to *Yes,* or the ratio of *No* to *Yes.* The investigation will focus on the first two: $\frac{\text{Number of } Yes}{\text{Total number}}$ and $\frac{\text{Number of } Yes}{\text{Number of } No}$.

INVESTIGATE

The answers to the questions will depend on the survey results. Here are sample results from a survey with 40 responses, 8 *Yes* and 32 *No,* on the first question.

Q3 32 *No* responses; 8 *Yes* responses

Q4 About $\frac{1}{5}$ of the ribbon is *Yes.*

Q5 The *Yes* bin is about one-fourth the size of the *No* bin.

Q6 a. $\frac{8}{40} = \frac{1}{5} = 0.2$ b. $\frac{32}{40} = \frac{4}{5} = 0.8$ c. $\frac{8}{32} = \frac{1}{4} = 0.25$
d. $\frac{32}{8} = \frac{4}{1} = 4.0$ e. $\frac{40}{32} = \frac{5}{4} = 1.25$ f. $\frac{40}{8} = \frac{5}{1} = 5.0$

Q7 Both $\frac{1}{5}$ and $\frac{1}{4}$ are ratios of *Yes* responses, but they compare the *Yes* response to different things. One-fifth of all responders thought that students should wear uniforms. One-fourth as many responders said *Yes* to uniforms as said *No* to uniforms. Some students may write sentences about the ratios that have *Yes* in the denominator.

Q8 The ribbon chart shows the ratio of *Yes* to total, and the bar chart shows the ratio of *Yes* to *No*.

6. The following answers are for data with 16 *Yes* and 24 *No* responses for the second question.

Q9 Any two of the ratios for *Response_2* answers can be included, along with sentences explaining their meaning.

$$\frac{\text{Number of } \textit{Yes}}{\text{Total number}} = \frac{2}{5}, \quad \frac{\text{Number of } \textit{No}}{\text{Total number}} = \frac{3}{5},$$
$$\frac{\text{Number of } \textit{Yes}}{\text{Number of } \textit{No}} = \frac{2}{3}, \quad \frac{\text{Number of } \textit{No}}{\text{Number of } \textit{Yes}} = \frac{3}{2},$$
$$\frac{\text{Total number}}{\text{Number of } \textit{No}} = \frac{5}{3}, \quad \frac{\text{Total number}}{\text{Number of } \textit{Yes}} = \frac{5}{2}$$

Q10 One way is to choose the first ratio and solve the proportion $\frac{\textit{Yes} \text{ responses in class}}{\text{Students in class}} = \frac{x}{\text{Students in school}}$. Students might choose different proportions for each question.

Q11 One way is to choose the second ratio and solve the proportion

$$\frac{\textit{No} \text{ responses in class}}{\text{Students in class}} = \frac{\textit{No} \text{ responses in your grade}}{\text{Students in school}}$$

EXPLORE MORE

1. Answers will depend on the data.
2. Answers depend on class data.
3. People cannot help thinking in terms of patterns, so any group they deliberately select is likely to have one or more things in common. Statisticians use randomness to reduce this bias. In a random selection, each subject has an equal chance of being selected. The selection is not based on any characteristics the group has in common. Randomness will not definitely eliminate all commonalities, but it is as fair as we can be. Any nonrandom selection process may shift the results of the survey one direction or another.

EXTENSIONS

1. Add the second attribute by dropping it in the middle of the ribbon chart of the first response. What can students say about this graph? What ratios appear here, and how are they interpreted?
2. If the two questions you asked are related (such as *Response_1* indicating opinions about longer classes and *Response_2* asking whether school should start earlier), students might look at and analyze the numbers that responded *Yes* to both, *No* to the first and *Yes* to the second, and so on.

Weighted Average—Swimmers

As part of a study on the impact of adding a swimming pool at the middle school, a student gathered these data on a sample of 200 people from her community, grouped by age (adult or teen) and by ability to swim (N = can't swim; Y = can swim).

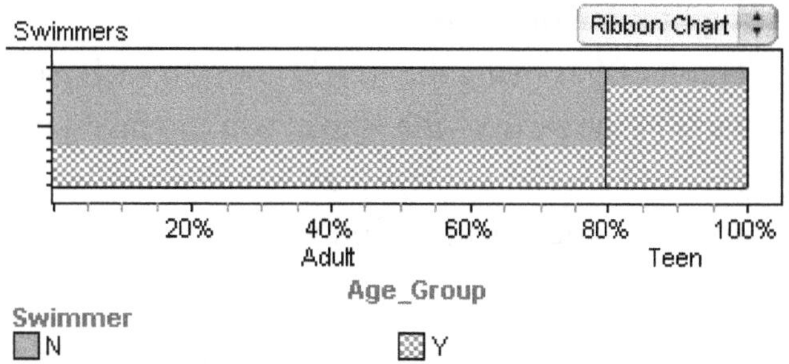

Q1 From the graph, estimate the percentage of adults who can't swim and the percentage of teens who can't swim.

Q2 Now estimate the percentage of the entire group who can't swim.

INVESTIGATE

1. To see the data for these 200 people, open **Swimmers.ftm.**

Summary

2. To see statistics that summarize a data set, drag a new summary table into the document.

3. Drag the *Swimmer* attribute from the case table to the summary table. Drop it over the arrow pointing down. Then drag the *Age_Group* attribute from the table and drop it over the arrow pointing to the right. Each box of the table now contains a number.

The meaning of the numbers appears below the summary table.

Q3 Three of the numbers appearing in the summary table are 107, 40, and 86. Use a complete sentence to describe what each number means.

Q4 By hand, calculate the percentage of the individuals who are teens.

4. You can have Fathom calculate the fraction of each category in the column. Select the summary table and choose **Summary | Add Formula.** Enter the single word columnProportion. (The word will change color when it is spelled correctly.)

You might need to drag a corner of the summary window to see the second number in each box and the description below the table.

Q5 Use complete sentences to describe the meaning of the table's numbers 0.825, 0.66875, and 0.57 (or their percentage equivalents).

Swimmers

		Age_Group		Row Summary
		Adult	Teen	
Swimmer	N	107 0.66875	7 0.175	114 0.57
	Y	53 0.33125	33 0.825	86 0.43
Column Summary		160 1	40 1	200 1

S1 = count( )
S2 = columnProportion

Q6 What is the average (mean) of the ratio of adults who can't swim to all adults and the ratio of teens who can't swim to all teens? Compare your answer with the information in the table that says 57% of people can't swim. Explain why the numbers are the same or different.

Is there a way to find the ratio for the full group from the ratios of the parts? You might think of the average you just found, $\frac{0.66875 + 0.175}{2}$, as $0.66875\left(\frac{1}{2}\right) + 0.175\left(\frac{1}{2}\right)$. Think of $\frac{1}{2}$ as being the *weight* given to each part. To get from the ratios of unequal parts to the ratio of the whole, you can give a weight of more than $\frac{1}{2}$ to the larger part and a weight of less than $\frac{1}{2}$ to the smaller part. The result is called a *weighted average.* Experiment to find what weights you should use for a weighted average in this study.

Slider

When two words have to be made into one word, programmers sometimes use CamelCase, using capital letters to help distinguish the words. You can use CamelCase rather than an underscore for your slider names.

5. To experiment with different values for the weights, drag a slider from the shelf to your document. Be sure the name *V1* is selected and enter a name for the slider, such as *Weight_for_Adults.* Pull down another slider and give it a name like *Weight_for_Teens.*

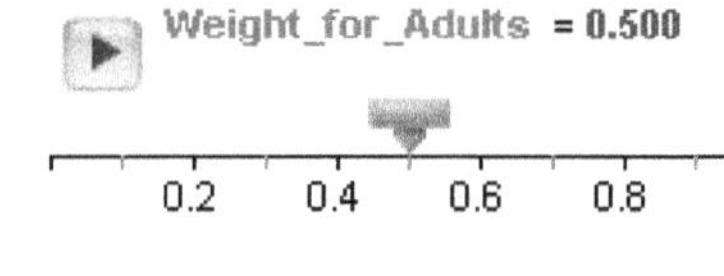

6. You want the slider values to be between 0 and 1, because two weights will add to 1. Double-click on a slider "thumb" to open the slider inspector. Set *Lower_* to 0 and *Upper_* to 1. Do the same for the other slider.

Inspect Slider

Properties

Property	Value
Weight_for_Adults	0.5
Max_updates_per_second	
Lower_	0
Upper_	1

7. You can also make a slider whose value depends on the values of other sliders. Drag down a third slider, and name it something like *Weighted_Ave.* Go to **Edit | Edit Formula** and enter the formula

0.66875 • Weight_for_Adults + 0.175 • Weight_for_Teens

Q7 Explain this formula. Where do the numbers 0.66875 and 0.175 come from?

8. Slide *Weight_for_Adults* and *Weight_for_Teens* until the value *Weighted_Ave* equals 0.57. There is more than one possible pair of answers.

Q8 What values did you find for *Weight_for_Adults* and *Weight_for_Teens?*

You might use 1 – *Weight_for_Teens.*

9. In Q4, you found the ratio of the number of teens to the number of subjects in the whole group. Now find the ratio of the number of adults to the number of subjects in the whole group. The sum of these ratios should be 1. Use the fact that the sum is 1 to write an equation expressing one weight ratio in terms of the other; use the slider names *Weight_for_Adults* and *Weight_for_Teens.* Enter the formula in the appropriate slider.

Q9 What formula did you use?

10. Adjust the slider that can still be moved independently until the weighted average of ratios of parts equals the ratio for the whole.

Q10 What weighted average gives you the ratio for the whole?

Q11 Restate in words, and explain, the equation for the weighted average.

EXPLORE MORE

1. The table also contains an *Age_Range* attribute. In the graph and in the summary chart, replace *Age_Group* with *Age_Range*. Find a formula for the weighted average of the three swimmer ratios to equal 0.57, the ratio in the summary table.
2. A student has a homework average of 80% and a test average of 95%. Her teacher gives grades based on a weighted average of these two averages. The student's weighted average is 90%. Another student has a homework average of 90% and a test average of 70%. What is his weighted average? Demonstrate this with sliders.

Weighted Average—Swimmers

Objective: Students will explore the ratios used in calculating a weighted average. Fathom calculates the ratios of two groups and the ratio for the total. Using sliders, students can determine the weights that make the weighted average of ratios of parts equal to the ratio for the total.

Student Audience: Pre-algebra, Algebra 1

Activity Time: 30–45 minutes

Setting: Paired/Individual Activity (use **Swimmers.ftm**)

Mathematics Prerequisites: Students can use ratios and average two numbers.

Fathom Prerequisites: Students can open a Fathom document.

Fathom Skills: Students will create and use formulas in two-way summary tables and use sliders with formulas.

Notes: Estimating the two ratios from the graph and recognizing the connection between the ratio and a percentage may be easy for some students but difficult for others. Encourage students to give complete answers to Q5–Q7 and Q11. Opportunities to explain what they are doing and the meaning of the result deepen students' understanding. As students share a variety of answers for Q8, ask how they can all be valid answers.

Q1 Answers should approximate $\frac{2}{3}$, or 0.67, for adults and $\frac{1}{6}$, or 0.17, for teens.

Q2 Some students may get 0.42 by averaging the averages rather than the actual ratio of 0.57. This question might cause some disagreement among students. Use discussion to spark interest in exploring this discrepancy.

INVESTIGATE

Q3 The group contains 107 adults who don't swim, 40 teens, and 86 people who do swim.

Q4 Twenty percent, or $\frac{1}{5}$, of the group are teens.

Q5 The ratio of teens who swim to all teens is 0.825; that is, 82.5% of teens are swimmers. Of the adults, 66.9% don't swim. The ratio of people who don't swim to the whole group is 0.57; equivalently, 57% of the group are not swimmers.

Q6 $\frac{0.66875 + 0.175}{2}$, or approximately 0.42. Students may have different ideas about why this average differs from 57%. Through discussion, elicit the idea that the discrepancy is due to having different numbers of teens (40) and adults (160). If there had been the same number of adults and teens among the 200 people, then the average would have given the ratio for the entire group. Because there are different numbers of adults and teens, the average of the ratios for the parts is different from the ratio for the whole. The true ratio (57%) is closer to the ratio for the larger group (67%) than to the ratio for teens (18%).

Q7 0.66875 and 0.175 are the ratios of parts from the summary table. These will be weighted differently to find the ratio for the whole.

Q8 Several answers are possible.

9. Eighty percent, or $\frac{4}{5}$, of the group are adults; $\frac{4}{5} + \frac{1}{5} = 1$.

   Weight_for_Adults + *Weight_for_Teens* = 1
   Weight_for_Adults = 1 − *Weight_for_Teens*

Q9 In the *Weight_for_Adults* slider, enter **1 – Weight_for_Teens**, or in the *Weight_for_Teens* slider, enter **1 – Weight_for_Adults**.

Q10 $0.66875 \cdot 0.8 + 0.175 \cdot 0.2 = 0.57$

Q11 Because adults make up 80% of the population, their ratio is multiplied by 0.80, and the teens' ratio is multiplied by 0.20. If you multiply the ratio for each group by the importance of that group (that is, the ratio of the number of people in that group compared with the total number), you get the overall ratio.

EXPLORE MORE

1. $0.175 \cdot 0.2 + 0.627 \cdot 0.55 + 0.76 \cdot 0.25 \approx 0.57$
2. $(0.80)\left(\frac{1}{3}\right) + (0.95)\left(\frac{2}{3}\right) \approx 0.90$, or 90%

   $(0.90)\left(\frac{1}{3}\right) + (0.70)\left(\frac{2}{3}\right) \approx 0.77$, or 77%

   One way is to set the slider formula on the weighted average slider to **0.8 • hw_Weight + 0.95 • (1 – hw_Weight)** and move *hw_Weight* to 0.333 to get the formula to equal 0.90. Then, without moving *hw_Weight,* set a formula to **0.9 • hw_Weight + 0.7 • (1 – hw_Weight)**.

Proportions—Veterans

Data from a sample of a population are often used to predict something about that population. For example, of 500 Iowa residents in a random sample taken from the 2000 U.S. census, 55 people had served in the military.

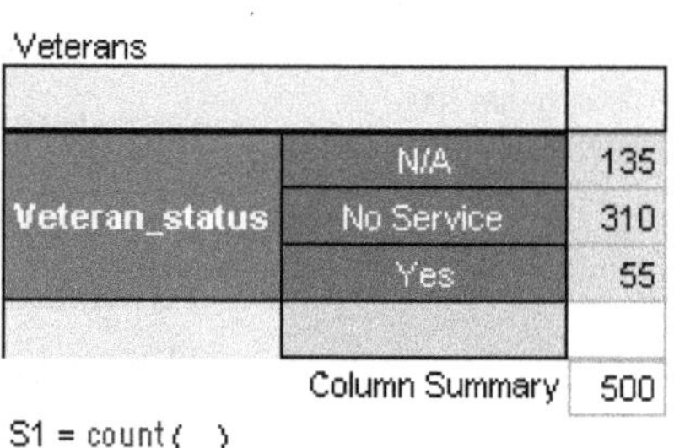

Veterans

Veteran_status	N/A	135
	No Service	310
	Yes	55
	Column Summary	500

S1 = count( )

The population of the state of Iowa in 2000 was 2,926,324. The sample might be used to estimate the number of veterans in Iowa. How good are estimates of the number of veterans based on this sample?

Q1 Set up a proportion stating that the ratio of veterans to the number of Iowans in the sample equals the ratio of Iowa veterans to the population of Iowa. A variable will represent Iowa veterans in your proportion. Solve this proportion to estimate the number of veterans in Iowa.

Q2 The U.S. population in 2000 was 281,421,906. Use the sample data to set up and solve a proportion for estimating the number of veterans in the United States. What percentage are veterans?

Q3 How accurate do you think these estimates might be? Explain.

INVESTIGATE

1. Open **Veterans.ftm.** You will see a collection (a box of balls) and a table and summary for Iowa data.

You must be connected to the Internet to download data.

2. Look at data from another state. Double-click on the collection to show its inspector and click on the last tab, **Microdata.** Under Choosing Cases, choose States. You will see a list of states, with an X in the box beside Iowa. Click on the X to remove it, then click on the box beside a different state. Click the **Download Data** button.

Inspect Veterans

Cases | Measures | Comments | Display | Categories | Microdata

Choosing Cases: Years, States, Metropolitan A, People/Housin, Sample Size; Attributes

☐ Hawaii
☐ Idaho
☐ Illinois
☐ Indiana
☒ Iowa
☐ Kansas
☐ Kentucky
☐ Louisiana
☐ Maine
☐ Maryland
☐ Massachusetts

Cases: 2000; Iowa; People Only - Attributes: Year; State; Age; Sex: Race: Marital status: Veteran Status

IPUMS: www.ipums.org

500 cases in collection | Download Data

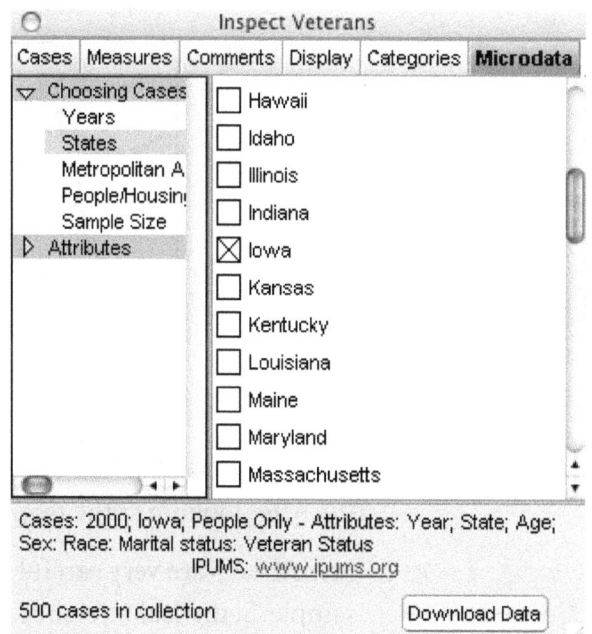

Q4 About how many veterans are in the state you chose?

The collection State Populations has 2000 census data.

Q5 According to the summary table, what is the ratio of veterans to residents of this new state?

Q6 If you used this ratio and the U.S. population (281,421,906), what would be the estimated number of veterans in the United States? What percentage is that? How accurate is this estimate? Explain.

Q7 Combine the information from Iowa and from your chosen state (giving you a sample of 1000 people) to give another estimate of the number of veterans in the United States. Is this estimate more accurate or less accurate than the estimate in Q5? Explain.

3. Return to the inspector and remove the X from your selected state. Now select a new state and download its data.

Q8 What is the ratio of veterans to residents in this sample? About how many veterans live in your new state?

Q9 Combine the information from Q7 with the information for your second state and Iowa (giving you a sample of 1500 people) and give a ratio of veterans and another estimate of veterans in the United States. Is this estimate more accurate or less accurate than your previous estimates? Why?

4. Look at the map, which shows how veterans are distributed across the United States. It shows why selecting some states might give an inaccurate estimate for the country.

Percentage of Population Who Are Veterans

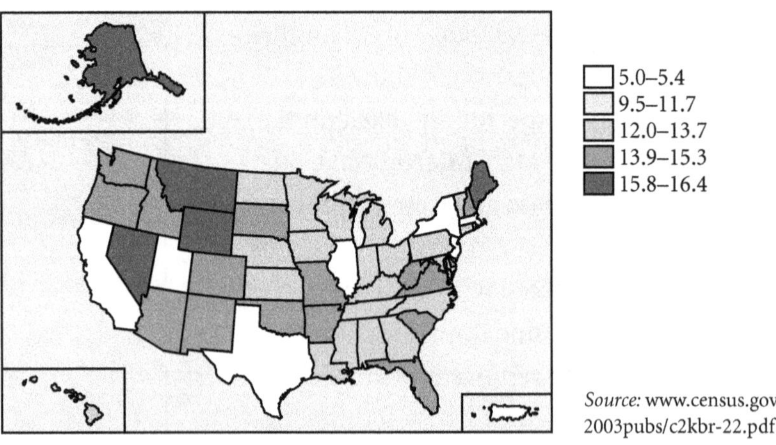

Source: www.census.gov/prod/2003pubs/c2kbr-22.pdf

Q10 After looking at the map, how accurate do you think your latest estimate is?

Statisticians are very careful about drawing inferences, or conclusions, from a sample. Statistical inference is only valid when the sample is chosen randomly. A sample of 50 from Nevada is likely to have more veterans than a sample of 50 from the entire United States.

Q11 Choose five states randomly and use this sample to estimate the percentage of veterans in the United States. Describe how you took your random sample and the estimate your sample predicts. How close do you come to the actual percentage of a little more than 9%?

EXPLORE MORE

1. On the **Microdata** panel of the collection inspector, pick a state or states. Then select Attributes: Person: Work: In Labor Force and collect a set of data so you can estimate the ratio of persons in the labor force. Use that number to estimate the number of U.S. residents in the labor force.

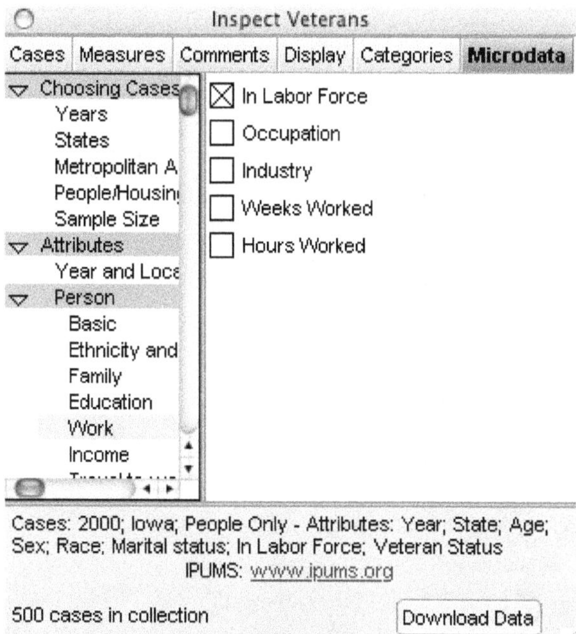

2. Choose another attribute from the **Microdata** panel and explore those data from several states. Describe the patterns you find.

Objective: Students will deepen their understanding of setting up and solving proportions. Fathom allows students to gather census microdata about any of the 50 states and make estimates about statistics for the entire country. This activity is also an informal look at how biased sampling can affect predictions and estimates.

Student Audience: Pre-algebra, Algebra 1

Activity Time: 20–30 minutes

Setting: Paired/Individual Activity (use **Veterans.ftm**)

Mathematics Prerequisites: Students can solve a proportion and read and understand a graph key.

Fathom Prerequisites: Students can use a summary table.

Fathom Skills: Students will use a collection inspector to collect microdata from a U.S. census.

Notes: Computers used for this activity must have access to the Internet for sampling U.S. census data. Samples of individual cases are drawn from the 2000 census data in steps 2 and 3. Because only one state can be sampled at a time, students need to make new proportions by adding numbers from two, three, then five samples. Students might use technology to help them combine information, or they might use a pencil and paper. If you have students who make a table in Fathom or on their calculators, you might ask them to present it to the class. As you visit groups, ask for oral explanations to Q6 and Q7; encourage different explanations and note those that the class could benefit from hearing. You might ask groups to put their answers to Q9 on a class table. How do these estimates compare? Doing the same for Q11 could lead to some interesting class discussion. You might ask students with good methods of taking a random sample of states to share their methods with the class.

Q1 $\frac{55}{500} = \frac{v}{2,926,324}$; the estimated number of veterans is 321,896.

Q2 $\frac{55}{500} = \frac{v}{281,421,906}$, so $v = 30,956,410$, or 11%

Q3 The estimate for Iowa might be fairly accurate, because it was a random sample of Iowans. The estimated number of veterans in the U.S. will be accurate only if Iowa represents the whole country.

INVESTIGATE

Q4–Q9 Answers will vary depending on the states and the samples. Sample answers below are for adding data from New York and Minnesota. State populations are in a separate collection in the Fathom document.

Q4 There are about 1,783,800 veterans in New York.

Q5 New York: $\frac{47}{500}$, or 9.4%

Q6 $\frac{47}{500} = \frac{v}{281,421,906}$; $v = 26,453,659$

This is somewhat lower than the answer using Iowa. Perhaps the true number is in between.

Q7 $\frac{102}{1000} = \frac{v}{281,421,906}$

$v = 28,705,034$, or about 10.2%

This estimate is probably more accurate, because there is a larger, more diverse sample.

Q8 Minnesota: $\frac{65}{500}$; Veterans in Minnesota:
$\frac{65}{500} = \frac{v}{28,705,034} = 3,731,654$

Q9 $\frac{167}{1500} = \frac{v}{281,421,906}$; $v = 31,331,639$

This is probably no more accurate than just Iowa. Minnesota and New York were on opposite extremes from Iowa.

Q10 Different samples give different estimates. Because not all states are equally represented in the military, selecting only one or two states will bias the estimate for the country.

Q11 In general, unbiased samples including more states give estimates closer to the 9.4% for the population, or about 26,454,000 veterans in the United States. Here are two good descriptions of the random sampling.

- Sample answer: We used a random number generator on our calculator to select numbers from 1 to 50 and used the case numbers from the population table to find the corresponding states.
- Sample answer: We used Fathom to take a random sample from the case table of State Populations.

EXPLORE MORE

1. The actual number in the labor force is about 142,600,000 people.
2. Answers will vary.

Dimensional Analysis—Fastest Animals

The book *Top 10 of Everything in 2004* lists the ten fastest animals with their speeds in meters per second. Some characteristics of these ten animals are given with measurement units. You will often find it useful to convert to different measurement units. Your goal is to find quick ways to do that conversion.

Q1 The fastest animal is the cheetah, with a speed of 29.1 meters per second (m/s). How fast do you think that is in miles per hour (mi/h)?

Q2 Is it possible to convert from m/s to mi/h in a single step? Explain why or why not.

INVESTIGATE

1. Fathom can convert among many common measurement units. To see how, start by opening the document **FastestAnimals.ftm.**
2. Select the case table. Display the units by choosing **Table | Show Units.**
3. Select the attribute *Speed* by clicking on the attribute name. Choose **Edit | Copy Attribute,** then immediately choose **Edit | Paste Attribute.**
4. In the units row of the new attribute, *Speed1* (to the left of *Speed*), change the unit to miles per hour.

Fathom abbreviates miles per hour as mph or mi/h.

Fastest Animals

	Name	Speed1	Speed
units		mi/h	m/s
1	Cheetah	65.1 mi/h	29.1 m/s
2	Pronghorn antelope	55.0 mi/h	24.6 m/s
3	Mongolian gazelle	50.1 mi/h	22.4 m/s
4	Springbok	50.1 mi/h	22.4 m/s
5	Grant's gazelle	47.0 mi/h	21.0 m/s
6	Thomson's gazelle	47.0 mi/h	21.0 m/s
7	Brown hare	45.0 mi/h	20.1 m/s
8	Horse	42.9 mi/h	19.2 m/s
9	Greyhound	42.1 mi/h	18.8 m/s
10	Red deer	42.1 mi/h	18.8 m/s

Q3 What is the speed of a cheetah in mi/h?

Q4 Which of these animals could outrun a car going 50 mi/h?

When converting from one unit to another, you can use the idea that multiplying by 1 does not change a number. One inch (in.) equals 2.54 centimeters (cm); so the fraction $\frac{2.54\text{ cm}}{1\text{ in.}}$ and the fraction $\frac{1\text{ in.}}{2.54\text{ cm}}$ both equal 1; they are unit fractions. To change 30 cm to inches, multiply 30 cm $\cdot \frac{1\text{ in.}}{2.54\text{ cm}}$, which is about 12 in. To change 50 in. to centimeters, multiply 50 in. $\cdot \frac{2.54\text{ cm}}{1\text{ in.}}$, which is about 127 cm. Products of unit fractions are like those of numeric fractions: the units in the numerator of one cancel the same units in the denominator of another. But how do you know that 1 in. equals 2.54 cm, as claimed above? Fortunately, Fathom can help you find conversion factors like this.

5. In the case table, to the right of all existing attributes, click on <*new*> to create a new attribute called *Change.*

Fastest Animals

	Change
units	meters
1	

6. Enter the units for this attribute as meters.

7. For the first data item in this column, enter 1 mi. When you press **Enter,** the value will change to the number of meters that are in 1 mi.

Q5 Based on this change, write two fractions that equal the number 1. (Remember to include the units.)

Q6 Which of these fractions would you multiply by $\frac{32\text{ m}}{\text{s}}$ to change to $\frac{\text{mi}}{\text{s}}$?

8. Change the units of your new attribute to seconds and enter the first value of 1 min. When you hit **Enter,** the 1 min changes to the equivalent measure in seconds.

Fastest Animals

	Change
units	seconds
1	1 min

Q7 How could you multiply $\frac{29.1\text{ m}}{\text{s}}$ by *two* fractions to change to $\frac{\text{mi}}{\text{min}}$? Remember that a unit you do not want in the final results must appear in both a numerator and a denominator in order to cancel.

Q8 How could you multiply $\frac{29.1\text{ m}}{\text{s}}$ by *three* fractions to change to $\frac{\text{mi}}{\text{h}}$?

With the table selected, choose **Table | Show Formulas.**

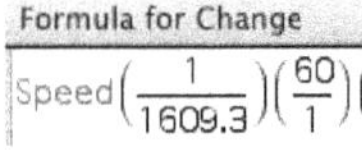

9. Select the units on *Change* and delete them. Also delete the units on *Speed* and *Speed1*. For the attribute *Change,* enter the formula to change the units that you developed in Q8 (without the unit labels), Speed $\left(\frac{?}{?}\right)\left(\frac{?}{?}\right)\left(\frac{?}{?}\right)$.

Q9 Compare the values of *Change* with the values in the other columns.

10. Create another new attribute named *Ratio.* Select the column, choose **Edit | Edit Formula,** and enter the formula Speed1/Speed.

Fastest Animals

	Ratio
units	
=	Speed1 / Speed

Q10 How do the values in the new column relate to the conversion you found in step 9? Explain.

EXPLORE MORE

1. Reenter the units in *Speed1* (mi/h). Next change the units to mach (the speed of sound). Then delete the units. How fast is the speed of sound in m/s? Explain how you found this value. How fast is the speed of sound in mi/h?

2. Use Fathom to determine the conversion from the speed of sound (mach) to the speed of light (lightspeed).

3. Drag a new graph from the shelf. From the case table, drag *Speed* to the graph's horizontal axis and *Speed1* to its vertical axis. Make sure that *Speed1* does not have units. Now choose **Graph | Add Movable Line** then **Graph | Lock Intercept At Zero.** Adjust the line to fit the data. How does the equation of this line relate to the conversion of *Speed* to *Speed1?*

Dimensional Analysis—Fastest Animals

Objective: Students will use ratios equal to one, called *unit ratios*, to convert units. As Fathom calculates the conversion factor, students will see validation of the conversion factor they found from dimensional analysis.

Student Audience: Pre-algebra, Algebra 1

Activity Time: 30–40 minutes

Setting: Paired/Individual Activity (use **Fastest Animals.ftm**)

Mathematics Prerequisites: Students understand units and can reduce fractions by dividing the numerator and denominator by the same quantity.

Fathom Prerequisites: Students can open Fathom documents and add new attributes and their own formulas.

Fathom Skills: Students will duplicate an attribute, change or delete attribute units, and add a line through (0, 0) to a scatter plot (Explore More 3).

Notes: The first two questions are for thought and need not be discussed beyond the pair or small group working together. Return to question Q2 at the end of the activity to make sure all students agree with the answer. As you visit groups, ask why. Why does $\frac{1\text{ mi}}{1609.34\text{ m}} = 1$? Why does $\frac{60\text{ s}}{1\text{ min}} = 1$? Why can you use a ratio that equals 1 (a unit ratio) to convert a measurement to new units? Why can you cancel units? How can you be sure you are using the correct ratios?

These data come from the book *The Top 10 of Everything in 2004;* publication information can be found on the **Comments** panel of the collection's inspector. The book is a good source of small data sets to study and graph.

Q1 About 65 mi/h

Q2 Yes; any units that can be converted in multiple steps can be converted in one step by combining all the steps into a single expression, a unit ratio.

INVESTIGATE

Q3 65.1 mi/h

Q4 The top two (cheetah and pronghorn antelope) would outrun the car, and the next two (Mongolian gazelle and springbok) would keep up with the car.

Q5 $\frac{1609.34\text{ m}}{1\text{ mi}}$ and $\frac{1\text{ mi}}{1609.34\text{ m}}$

Q6 $\frac{1\text{ mi}}{1609.34\text{ m}}$

Q7 $\left(\frac{29.1\text{ m}}{\text{s}}\right)\left(\frac{1\text{ mi}}{1609.34\text{ m}}\right)\left(\frac{60\text{ s}}{1\text{ min}}\right) = \frac{1.085\text{ mi}}{\text{min}}$

Q8 $\left(\frac{29.1\text{ m}}{\text{s}}\right)\left(\frac{1\text{ mi}}{1609.34\text{ m}}\right)\left(\frac{60\text{ s}}{1\text{ min}}\right)\left(\frac{60\text{ min}}{1\text{ h}}\right) = \frac{65.095\text{ mi}}{\text{h}}$

Q9 The values of the *Change* attribute now match those of *Speed1*. With the Change column selected, choose **Table | Format Attribute** and set the attribute to a fixed decimal of one decimal place. Because of rounding, one of the measurements is slightly off.

Q10 This column is the conversion factor for m/s to mi/h. It is also the product of $\left(\frac{1\text{ mi}}{1609.34\text{ m}}\right)\left(\frac{60\text{ s}}{1\text{ min}}\right)\left(\frac{60\text{ min}}{1\text{ h}}\right)$.

EXPLORE MORE

1. 1 mach = 331.46 m/s = 741.455 mi/h
2. 1 lightspeed = 904,460 mach
3. The slope of the line is the conversion factor that you have found. (The intercept of the line should be 0.)

Proportions—Squirrel Population

To estimate populations in the wild, naturalists use a technique called *capture-recapture.* They capture and tag some of a population before releasing them. After giving the tagged animals some time to mix in with the others, they make another round of captures. They carefully note how many of those captured in the second round were tagged in the first round. They get data like these.

Squirrels Caught	
Both first and second	15
First only	270
Second only	153
Total	438

Q1 How many squirrels were captured and tagged in the first round? How many were captured in the second round? How many caught in the second round were already tagged?

Naturalists calculate the total population using a proportion that equates the ratio of the number of squirrels caught in the second capture to the number of those second-capture squirrels who are tagged with the ratio of the total population to the total number who were tagged on the first round.

Q2 What proportion that uses the three values from Q1 and a variable for the number in the total population can be used to estimate the population? What is the estimate of the total population?

INVESTIGATE

In this investigation, you will explore how naturalists decide how many animals to capture in each round.

1. Open the document **Squirrels.ftm,** which contains a collection consisting of the population of squirrels.

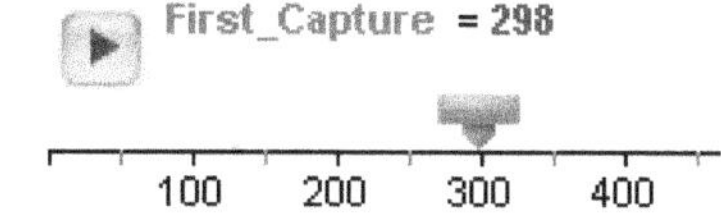

2. There is a slider named *First_Capture,* which allows you to capture up to 500 squirrels in the first round. Set this slider to something near 300.
3. There is a second slider named *Second_Capture,* which allows you to capture up to 500 squirrels in the second round. Set this slider, also, to something near 300.

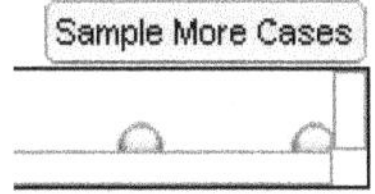

4. The first and second capture will happen each time you click on the **Sample More Cases** button. Do that now. The results of the second capture appear in the graph and the summary box.

Q3 Use each of the numbers in the summary box to write a sentence about what happened in the second capture.

Q4 Write a proportion using this information, the size of the first capture, and the unknown size of the full population. Then solve this proportion to predict the size of the squirrel population.

Q5 Compare your result with those of others around you. How satisfied are you with your prediction?

5. Set both of the sliders to only about 100 squirrels, then sample more cases.

Q6 Calculate the population size based on these values. Compare these results with those of others around you or repeat the simulation and compare the results.

Click on the Second Capture collection to open the inspector.

6. You can build a population calculator. Open the inspector for the second capture collection and click on the tab for **Measures.** Enter a new measure named Population. Rewrite this equation so that *Population* is alone on the left side: $\frac{Population}{First\_Capture} = \frac{Second\_Capture}{\text{count}(Tagged = \text{“Yes”})}$. Enter that formula.

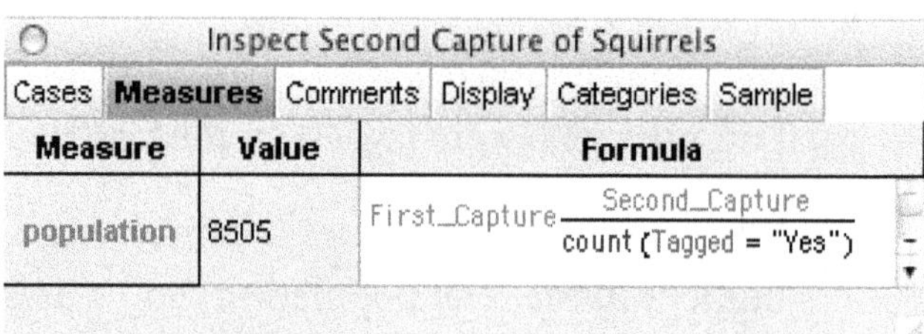

7. Select the Second Capture **Collection | Collect Measures.**

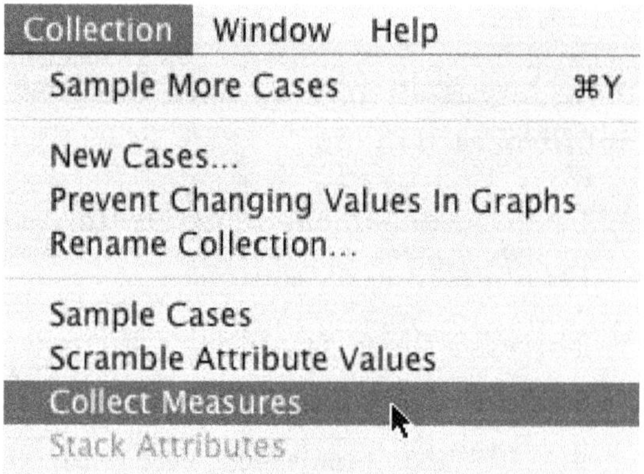

This will create another collection of population estimates. Each estimate is based on a new simulation of the capture-recapture procedure.

Highlight the Measures collection and drag a table from the shelf.

8. Create a case table of these measures and a dot plot of *Population.*

Measures from Second Capture of Squirrels

	Population	<new>
1	9922.5	
2	6615	
3	3969	
4	5412.27	
5	8505	

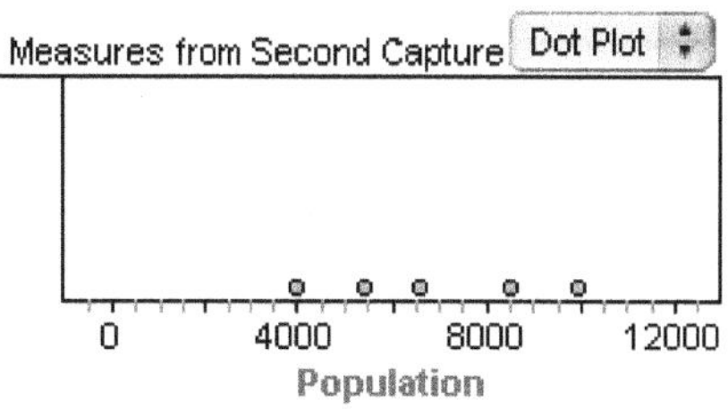

Q7 Interpret the results of this graph.

Once you understand where the measures are coming from, uncheck the **Animation on** box.

9. Open the inspector for the Measures collection; under the **Collect Measures** tab, set it to replace the existing cases and collect 20 measures. Reset one or both of the sliders, then collect more measures.

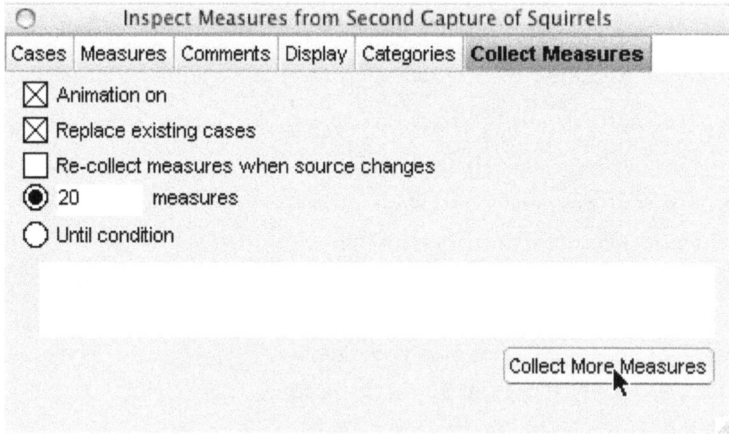

Q8 After exploring different sample sizes, what recommendations would you make about the sample size for an accurate estimate of the population?

Imagine that each capture costs $10 and the budget for the capture-recapture project is $6000.

10. Use your sliders to discover if it is better to make a large capture first or second or the same size each time. Remember that you are only going to capture a total of 600 squirrels.

Q9 What is your recommendation to the capture team?

EXPLORE MORE

IQR = Q3 − Q1. Here, Q3 is the third quartile and Q1 is the first quartile.

Calculate the interquartile range (IQR) of the 20 simulations using 600 captures total on each simulation. Then calculate the IQRs for simulations using 200, 400, 800, and 1000 captures. Summarize what you found.

Objective: As students simulate several capture-recapture scenarios, they will use ratios and proportions to make estimates. Students will see the effect of different scenarios on the accuracy of the estimate.

Student Audience: Algebra 1

Activity Time: 30–45 minutes

Setting: Paired/Individual Activity or Whole-Class Presentation (use **Squirrels.ftm** for either setting)

Mathematics Prerequisites: Students can set up and solve a proportion, and they understand what sampling is.

Fathom Prerequisites: Students can open a Fathom document, read a summary table, work with sliders, and create a dot plot.

Fathom Skills: Students will take samples from a collection and create and collect measures.

Notes: As you talk with students, help them understand two powerful tools of Fathom—the ability to collect samples and to collect measures on samples. Collecting samples involves cases selected at random from the original collection and assembled into a new collection. This sample collection will be a subset of the original. The original collection represents the entire population of squirrels; some squirrels were tagged in the first capture (the number depends on the value in the *First_Capture* slider). When Fathom collects a sample of Second Capture of Squirrels, it is simulating the second capture, which will have some tagged squirrels and many untagged squirrels. To estimate the population of squirrels, assume that the ratio of tagged to total sample in the second capture is the same as the ratio of total tagged (squirrels in the first capture) to the population.

A measure is a calculation based on the collection. For example, you could create a measure that counted the tagged squirrels in the collection. Any summary statistic, or any formula based on the summary statistic, can be made into a measure. After students have solved the proportion for the population, they are asked to turn this formula into a measure. When Fathom collects measures, it re-collects the sample, calculates the formula created for each measure, and then builds a new collection from the results of the calculation. The attributes in this new collection are the measures, and the cases each represent a measure taken from a different random sample.

For a Presentation: Using this activity as a presentation gives you a chance to make sure students understand what is happening when samples are taken and when measures of the samples are calculated. The animation helps students see that the sample comes from the Squirrels collection into the Second Capture of Squirrels collection and the measures come from that collection to the Measures from Second Capture of Squirrels. To collect large samples, turn off the animation. Take several samples for Q3 and Q4 and ask students to compare the results. Before taking samples or collecting measures, ask students what they think will happen. Building a graph like the one that answers Explore More gives several opportunities to ask the what-do-you-think question, which can keep students engaged.

Q1 In the first round, 285 (270 + 15) squirrels were captured and tagged. In the second round, 168 (153 + 15) were captured, and 15 of those had been tagged in the first round.

Q2 $\frac{\text{Total caught in second round}}{\text{Caught both first and second}} = \frac{\textit{Population}}{\text{Total caught in first round}}$; $\frac{168}{15} = \frac{\textit{Population}}{285}$; the estimated population is 3192 squirrels.

INVESTIGATE

Q3 Answers will vary. Sample answer: There were 305 squirrels caught—285 of them were caught for the first time, and 20 of them had already been caught and tagged in the first round of captures.

Q4 Answers in the form $\frac{\textit{Second\_Capture}}{\text{Number tagged in Second Capture}} = \frac{\textit{Population}}{\textit{First\_Capture}}$ will vary. Sample answer: $\frac{305}{20} = \frac{\textit{Population}}{298}$, *Population* = 4544 squirrels.

Q5 There will be a variety of estimates.

Q6 These results will likely be more varied than those for Q4. Sample answer: $\frac{103}{1} = \frac{\textit{Population}}{100}$, *Population* = 10,300 squirrels.

Q7 Each dot represents one estimate for the population's size based on one simulation of capture-recapture. The graph shows the spread in the estimates.

9. Measures can be collected faster with animation off, but watching the animation gives students a sense of where the numbers are coming from.

Q8 Most students will find that for large catches, there is less spread in the estimates of the population. Large samples provide better estimates. Some student results may not show this.

Q9 In general, the least spread occurs when the two samples are about the same size. Some student results may not show this.

EXPLORE MORE

As the sample size increases, the IQR decreases (accuracy increases). The decrease is not linear. The improved accuracy going from 600 to 800 is much less than the improvement from 400 to 600.

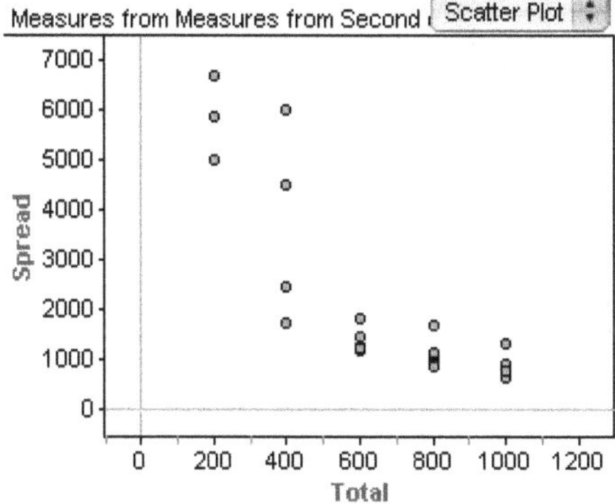

This graph was made by creating the collection Measures from Measures from Second Capture of Squirrels. Three measures were collected—*Spread,* with the formula iqr(Population); *Second_Capture,* with the formula Second_Capture; and *Total,* with the formula First_Capture + Second_Capture. To collect the measures, clear any data from Measures from Measures from Second Capture. Set the sliders to 100 and 100 and collect five measures. Reset the sliders for 200 and 200 and add five more measures. Repeat for 300 and 300, 400 and 400, and 500 and 500, giving 25 cases.

2

Exploring Linear Equations

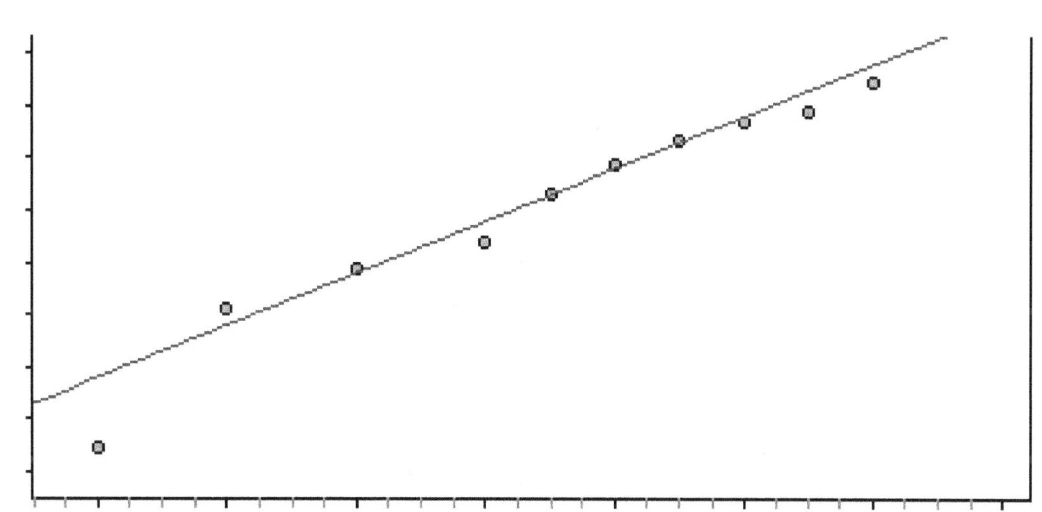

Slope—Runners

At the start of a long race, runners try to set a constant pace that they can maintain throughout the race. In this activity, you'll explore the graphs of data from runners.

Q1 Imagine a graph showing the distance a runner has traveled versus the time he has been running. What sort of pattern would you expect?

INVESTIGATE

1. Open **Runners.ftm.** You'll see two scatter plots. In the first, a machine has recorded locations (in meters) of three racers each second for 5 s (starting at 0 s).

Q2 Move the cursor onto the point that represents Ashley's distance after 1 s. Notice that the coordinates of this point, (1 s, 3.9000 m), appear in the status bar at the lower left corner of the Fathom window. What does this information tell you?

Q3 Move the cursor to the point that represents Ashley's distance after 5 s. Use the coordinates to determine how far Ashley traveled in 5 s.

2. Choose **Preferences** from the **Edit** menu (Windows) or from the **Fathom** menu (Macintosh). Choose $y = a + bx$ for the Linear Equation Form.

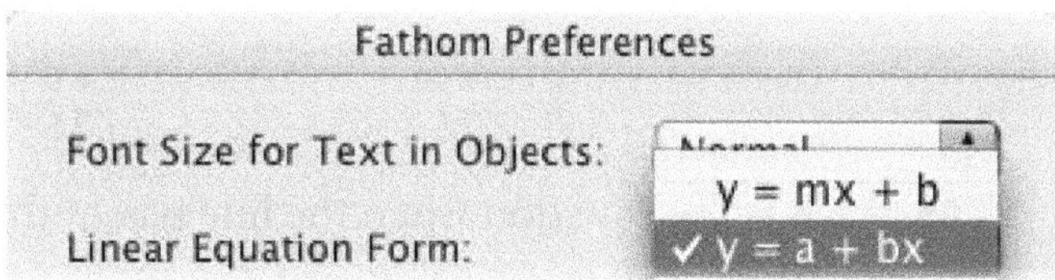

3. Click on the first scatter plot. From the **Graph** menu, choose **Add Movable Line.** A line will appear on the scatter plot. In the **Graph** menu, make sure **Lock Intercept At Zero** is turned on.

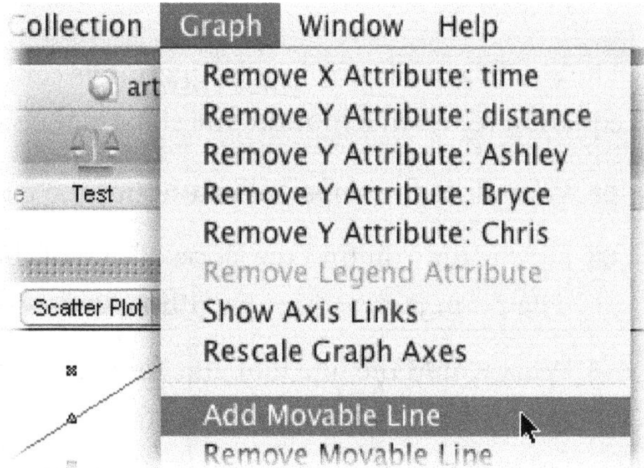

The rotation symbol looks like this.

4. Your first goal is to adjust the movable line until it passes through Ashley's points (the solid squares). When you put your cursor over the line, the cursor will turn into a rotation symbol. Drag the line to rotate it. Below the graph, an equation of the line (*distance* = 0 m + . . .) changes.

Q4 What is the equation when the line passes through the origin and the point representing Ashley's distance at 1 s?

5. Rotate the line to pass through the origin and the point representing Bryce's distance at 1 s.

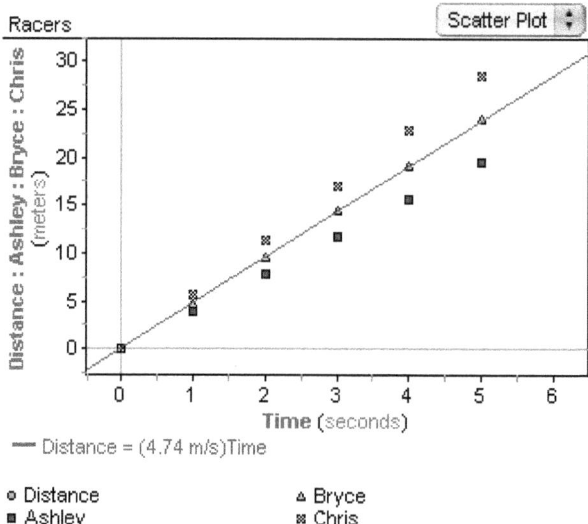

Q5 What is the equation when the line passes through the origin and the point representing Bryce's distance at 1 s?

Q6 What is the equation when the line passes through the origin and the point representing Chris's distance at 1 s?

Q7 How could you tell from the equations the distances the runners traveled in the first second?

The coefficient of time in the equation is called the *slope* of the line. When the equation represents motion, the slope is the same as the speed (in m/s for this case).

Q8 What does the slope indicate about the graph of the line?

Q9 How could you find the slope of one of these lines from the coordinates of the data point it goes through at time 2 s? At time 4 s?

Q10 What would the line look like for someone who ran faster than Chris? Slower than Ashley? What would the coefficient of *time* look like for someone who ran faster than Chris? Slower than Ashley?

After the race, the machine started accidentally and made two records of the locations of two children running in different directions along the track. This is the second scatter plot in the Fathom document.

6. Select this graph and add a movable line. Because the origin isn't showing on this graph, you won't lock the intercept at zero.

When you put your cursor over the middle of the line, it turns into an up-and-down arrow. If you click on that part of the line, you can drag the line up and down.

7. Adjust the movable line to pass through Dale's data points (the squares). You can rotate the line as before, or because it is not locked at zero, you can drag it up and down. The equation for the line changes as you move the line.

Q11 Dale's speed shows up in the equation as the slope of the line. What was Dale's speed?

8. Adjust the movable line to pass through Eddie's data points. (You'll need to do some extreme rotating.)

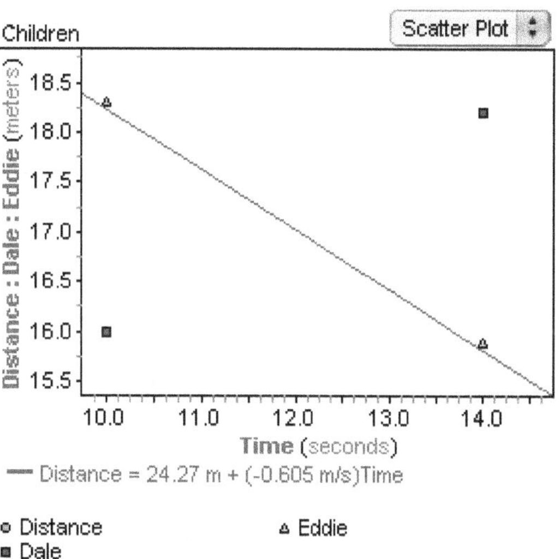

Q12 What was Eddie's speed? What does it mean that the slope of this line is negative?

Q13 In general, how can you find the slope of a line through two given points?

Q14 An equation of Eddie's line is approximately

$$distance = 24.3 + (-0.604)\ time$$

What does the first number (the one that's not a coefficient) represent in all the equations you've considered?

Slope—Runners

Activity Notes

Objective: Students will come to understand slope as a constant speed and as the steepness of a line. Students will learn that a line with a negative slope drops from left to right, and they will calculate the slope of a line by dividing the vertical change by the horizontal change (from zero).

Student Audience: Pre-algebra, Algebra 1

Activity Time: 25–35 minutes

Setting: Paired/Individual Activity or Whole-Class Presentation (use **Runners.ftm** for either setting)

Mathematics Prerequisites: Students can read the coefficient of a variable from an equation and understand coordinates of points in the rectangular coordinate system.

Fathom Prerequisites: Students can start Fathom and open a document.

Fathom Skills: Students will learn how to set the linear equation form, add a movable line, lock the line's intercept at zero, adjust the line, and read the equation.

Notes: Many students are attracted to the racing idea and may want to discuss the reasonableness of the speeds involved. There may also be questions about how constant the speeds can be if time 0 means when the runners begin. You can use the activity to bring out that slope will always be some kind of rate (though not always a speed) and will always have units such as meters per second, miles per gallon, or degrees Fahrenheit per foot below the surface.

As you circulate, ask students to describe how the equation for the line changes as they rotate the line. Ask further "what if" questions, such as those in Q10. Give students a chance to share with the whole class their answers and the reasoning behind their answers. Select different pairs to share their answers and thinking for Q7–Q13.

For a Presentation: You might use this activity to introduce Fathom and some of its features. To deepen students' understanding of the relationship between the equation and the line representing that equation, ask several students for answers to Q7–Q10 and Q13.

Q1 The points should lie in a line, because if the speed is constant, then time and distance should be proportional.

INVESTIGATE

Q2 Ashley ran 3.9 m the first second.

Q3 Ashley ran 19.5 m in 5 s.

Q4 Answers should be approximately *distance* = (3.9 m/s)*time.*

Q5 Answers should be approximately *distance* = (4.8 m/s)*time.*

Q6 Answers will vary. They should be approximately *distance* = (5.7 m/s)*time.*

Q7 The distances after 1 s are the coefficients of *time* in the equations.

You might reiterate that this number is called the *slope* and add that it's also the *rate.* The term *slope* applies to the graph because it describes the steepness. The term *rate* applies to the problem situation. In this case, it is a rate of meters per second, which is a speed.

Q8 The slope represents the steepness of the line.

Q9 Divide the distance coordinate (the change in distance from 0) by 2, which is the change in time during the first two seconds. Divide the distance coordinate by 4.

Q10 The line for a faster runner would be steeper; for a slower runner, the line would be less steep, or flatter. The coefficient (slope) would be larger for a faster runner and smaller for a slower runner.

Q11 Answers will be close to Dale's actual speed of 0.55 m/s.

Q12 Answers will be close to Eddie's actual speed of −0.6 m/s. Eddie moved in the direction opposite to the other runner's direction. The line is dropping from left to right.

Q13 Explanations should be equivalent to dividing the change in vertical coordinates by the change in horizontal coordinates. Students may say that they're dividing the vertical coordinate of a point not on the vertical axis by the horizontal coordinate of that point, because they may only be considering the case where one point is (0, 0).

Q14 The constant term is the y-intercept—the point where the line crosses the vertical axis. In these examples, it is the position of the runner when the time was 0. It can be thought of as the height from which a point rises or falls as it moves from 0 to the right along the line.

EXTENSION

Students can make their own measurements, enter them into a Fathom case table, and look for lines of fit that will allow them to make predictions. They might measure motion, change in temperature, or other rates.

Lines of Fit—Women's High Jump

Between 1972 and 1987, the world records for women's high jump increased consistently. The data points don't fit exactly along a line, but they do show a linear trend. By 2005, the record set in 1987 had still not been broken. Your goal is to find a line that fits the data fairly well so you can estimate what might have been the world record today had the trend continued.

INVESTIGATE

1. To see the women's high jump world records, open **High Jump Women.ftm.**

Click on the dot representing the most recent record and read from the status bar at the bottom of the Fathom window the record holder's name (year, record).

Q1 For the data shown, what is the record, when was it set, and who set it?

Q2 How much higher did the 1987 record holder jump than the high jump record set in 1971?

Q3 What is the horizontal difference between the first and last data point?

Q4 How much would a line through the first and last data point rise each year?

The rate, or rise each year, is the *slope* of the line.

Make sure that the Linear Equation Form is set to $y = a + bx$ (on Macintosh, choose **Fathom | Preferences;** on Windows, choose **Edit | Preferences**).

2. Choose **Add Movable Line** from the **Graph** menu. Adjust the line until it passes through the first and last data points.

Q5 How close is the coefficient of the year given in the equation of the movable line to the slope you calculated in Q4? Explain any differences.

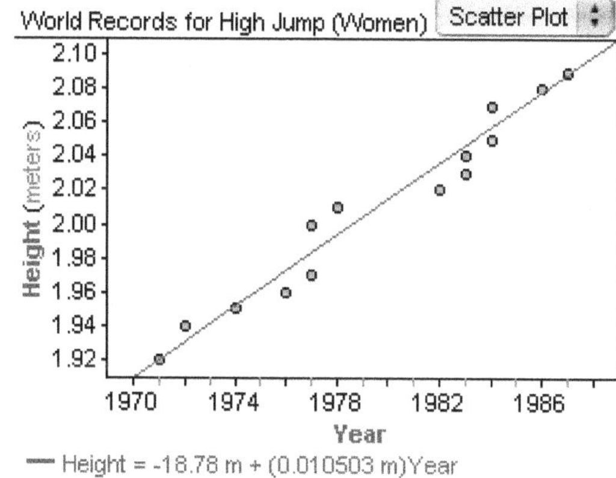

3. Adjust the movable line until it goes through any two data points and seems to follow the trend in the data with about as many points above the line as below.

Q6 What two data points does the line pass through?

Q7 What is the slope of the line between those points?

Q8 Use the line of fit between the data points you chose to predict what the record would have been in 2005 if the trend from 1972 to 1987 had continued?

By 2005, no one had broken the record of 2.09; in this case, it is not valid to extrapolate into the future. What about extrapolating into the past?

Q9 What would your current trend line predict for 1790? Does that make sense?

Q10 When do you think the world record might be broken? Explain.

EXPLORE MORE

Select the collection, drag a table from the shelf, add 2004 and 2.08 to the table, select the table, drag a graph from the shelf, drag *Time* to the horizontal axis, and drag *Height* to the vertical axis.

1. The Olympic record was set in 2004 at 2.08, slightly below the world record set in 1987. Add the point (2004, 2.08) to your data set. Describe a line of fit for the new data set.

2. In Q4, you calculated the slope of the line that went through the first and last points, but a line could have been drawn between other pairs of points. Some pairs of points would give a better line of fit than others. To randomly select pairs of points, create a sample collection from the World Records collection.

   Drag a new collection from the shelf and drop the name of the original collection on the new collection to create the collection Sample of World Records for High Jump (Women). You may rename your sample collection Two Points. (Click on the name to change it.)

   Double-click on the new collection to open its inspector; choose the **Sample** panel, and make the sample of size 2. Make sure With Replacement is not checked and Replace Existing Cases is checked.

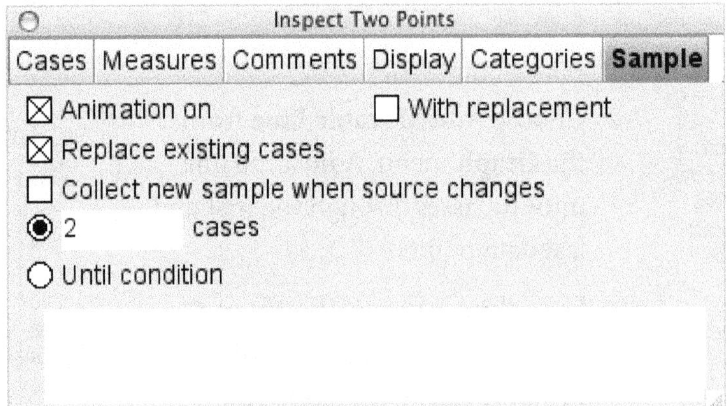

   Drag the attributes *Year* and *Height* from the **Cases** panel of the sample's inspector.

   Drag a new graph from the shelf. Drag *Year* to the horizontal axis and *Height* to the vertical axis to make a scatter plot. Change the scatter plot to a **Line Scatter Plot.**

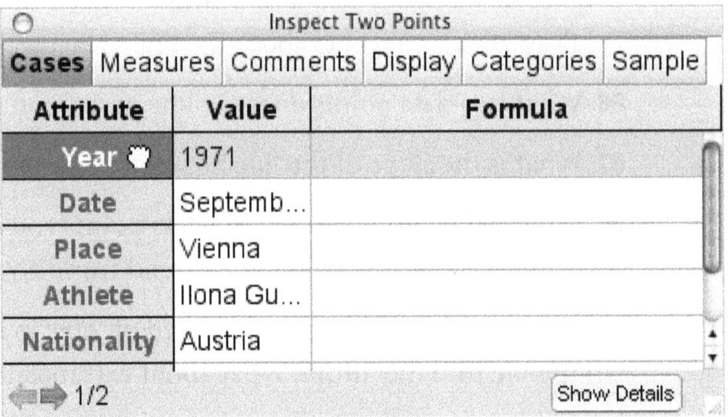

The **Sample More Cases** button is in the inspector on the **Sample** panel, or you can open the sample collection and use that **Sample More Cases** button.

Two Points | Sample More Cases

Click **Sample More Cases** until you get a line that seems to be about the same as the movable line in your first graph. How do the slopes compare?

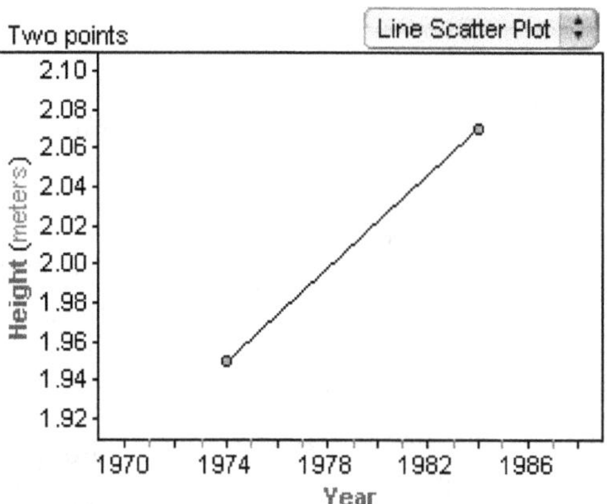

Click the **Measures** tab in the inspector to see the formulas used for the measures.

3. You can collect the measures from samples of two points, such as those you found in Explore More 1. In this Explore More, you will look at the slope of the line between the two points in each sample to see how close those measures of slope are to the slope of your line of fit. Open **High Jump Samples.ftm** and collect more measures. Describe the measures, what you see happening, and the results.

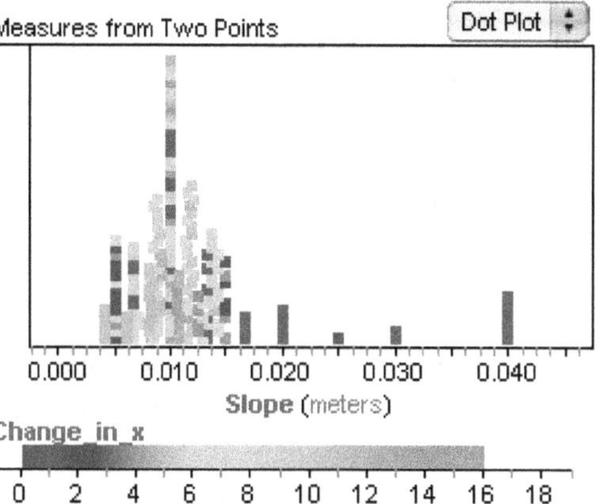

Inspect Two Points

Cases | Measures | Comments | Display | Categories | Sample

Measure	Value	Formula
Change_in_x	12	max (Year) − min (Year)
Change_in_y	0.11 m	max (Height) − min (Height)
Slope	0.00916...	$\frac{\text{Change\_in\_y}}{\text{Change\_in\_x}}$
<new>		

Objective: Using Fathom's movable line, students will identify a line of fit between two points. They will see that the best lines of fit are usually between points that are fairly far apart.

Student Audience: Pre-algebra, Algebra 1

Activity Time: 20–25 minutes

Setting: Paired/Individual Activity (use **High Jump Women.ftm**; use **High Jump Samples.ftm** for Explore More 3)

Mathematics Prerequisites: Students have some experience with linear equations, and they have general equation-solving and graph-reading skills.

Fathom Prerequisites: Students can open Fathom files and add movable lines to graphs. For Explore More, they should be able to sample cases and make a line scatter plot.

Notes: This activity, particularly Q8–Q10, can begin a good discussion of extrapolation. As you circulate and watch students complete step 3 and Q6–Q8, ask them to explain why they chose those two points.

As students share with the whole class their answers and their thinking, you want them to get a sense that when looking for a good line of fit, it's best to use representative data points that are fairly far apart. You might extend the discussion by asking students for examples of situations in which extrapolation might work. How might a person be able to determine whether a trend will continue?

The Explore More extensions require advanced Fathom skills. You might use Explore More 1 as a demonstration of the relationship among the collection, its table data, and a graph of those data. Presentations of Explore More 2 and 3 can help students visualize slope.

If Explore More 2 and 3 are students' first exposure to samples and measures, help them understand what is happening. When a sample is taken, cases (here, each case is a point) are drawn from a collection at random. In this instance, the sample is two points, and the line graph allows students to see the line between those points. Measures can be taken of something that has been sampled. In **High Jump Samples.ftm**, Fathom measures the slope of the line between the two points in the sample.

Ask students to explore and explain the formulas on the **Measures** panel of the Two Points inspector. Will these formulas always work? What formulas would always work? Was there a reason these general formulas were not used? Also, ask about the Measures from Two Points graph. Bring out that the colors of the dots indicate the difference in the x-coordinates of the two randomly chosen points, and the position of the dot on the graph tells something about the ratio $\frac{\Delta y}{\Delta x}$, or the slope. When the change in x is small, the color is blue; when the change in x is small and the change in y is relatively large, the slope is high. The colors were created by first making a dot plot of slope and then dropping *Change_in_x* in the center of the graph. Do the colors help? Do they help explain the outliers?

INVESTIGATE

Q1 Stefka Kostadinova set the record, 2.09 m, in 1987.

Q2 $2.09 - 1.92 = 0.17$ m; this is the vertical difference; the change in y.

Q3 $1987 - 1971 = 16$ yr; this is the change in x.

Q4 $\frac{0.17\text{ m}}{16\text{ yr}} \approx 0.0106\,\frac{\text{m}}{\text{yr}}$

Q5 Answers will vary, but the two slopes (m/yr) should be fairly close.

Q6 Answers will vary.

Encourage students to adjust the trend line so there are about the same number of data points above the line as below it.

Q7 Answers will be near the slope of 0.01 m/yr.

Q8 Predictions will vary. The record would probably have been 2.3 m or more.

Q9 A record jump of near 0 m in 1790, or negative distances before that, makes no sense.

Q10 There have been other long periods where no record was set, then finally the old record was shattered completely and patterns continued. This happens, in part, due to sudden changes in equipment, training, or sport popularity. Performance-enhancing drugs, used before they were illegal and tested for, could make some old records hard to break.

EXPLORE MORE

1. The new line might have a slope of about 0.005. It might go very near the points for 1982 and 1983, with six of the other points above the line and six below it.
2. Answers will vary, but slopes should be about 0.01 m/yr.
3. As a sample is taken, the measure of the slope is added to the graph. Most of the samples have slopes near 0.01 m/yr. Colors on the graph indicate the change in *x*. A few samples where the change in *x* is small have a large slope, because the change in *y* was large.

If students explore the formulas in the Measures from Two Points collection, they will see that the change in *x* and the change in *y* are determined by calculating the maximum minus the minimum for each coordinate. This formula will only work if the slope is always positive.

EXTENSION

Ask students to find out whether the record set in 1987 has yet been broken. If it has, add that point to the scatter plot. How does that change student estimates for the line of fit?

Intercept Form—Hot Dogs

Takeru Kobayashi won the 2004 hot-dog–eating contest by eating 53.5 hot dogs in 12 minutes. Your goal is to explore one way of competing with him.

Q1 How many hot dogs would you normally eat in 12 minutes?

Q2 At what rate was Takeru eating hot dogs?

INVESTIGATE

Edit | Preferences (Win); **Fathom | Preferences** (Mac)

1. Open a new Fathom file. Make sure that the Linear Equation Form is $y = a + bx$ in **Preferences.**

Choose **Table | Show Units** and **Table | Show Formulas.** To add cases, select the collection and choose **Collection | Add Cases.**

2. Drag a case table from the shelf and make attributes for *Time* and *Hot_dogs* to represent the number of hot dogs eaten. Add 13 new cases, which will be 13 consecutive minutes of contest time. Make the units for *Time* minutes and the units for the number of *Hot_dogs* eaten hd.

3. As a formula for *Time*, enter caseIndex. The term *caseIndex* refers to the numbers in the far left column, which count cases. Make a recursive formula for *Hot_dogs:* prev(Hot_dogs) + 4.46 hd.

Hot dogs

	Time	Hot_dogs
units	minutes	hd
=	caseIndex	prev(Hot_dogs) + 4.46 hd
1	1 min	4.46 hd
2	2 min	8.92 hd
3	3 min	13.38 hd
4	4 min	17.84 hd
5	5 min	22.3 hd
6	6 min	26.76 hd
7	7 min	31.22 hd

To rotate the line, move the cursor over the line until you get a rotation symbol.

4. Drag a new graph from the shelf. Drag the *Time* attribute to the horizontal axis, then drag the *Hot_dogs* attribute to the vertical axis. The graph turns into a scatter plot. Each dot has two coordinates: (*Time, Hot_dogs*). Choose **Graph | Add Movable Line.** Adjust the line so it goes through the data points.

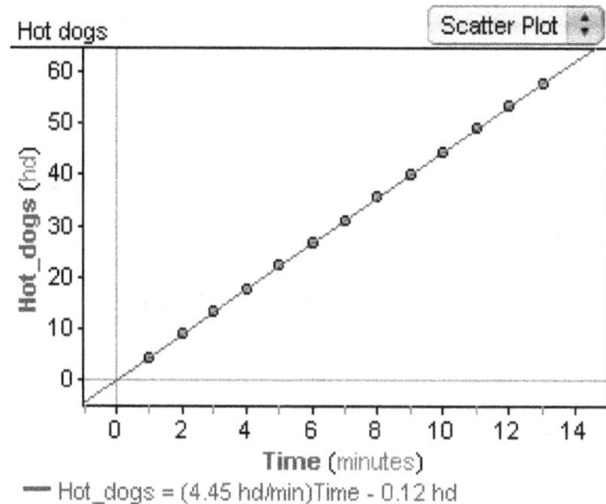

If you click **OK** before you feel that you have a good enough equation, double-click on the formula to return to the formula editor.

5. To determine how many hot dogs Takeru had eaten at any time, you'll plot a fixed line that lies on top of the movable line. Select the graph. From the **Graph** menu, choose **Plot Function.** You'll see the formula editor. The left side of the line's equation, *Hot_dogs,* is already given. Type in the right side of the line's equation and click **Apply.** Be sure to include units.

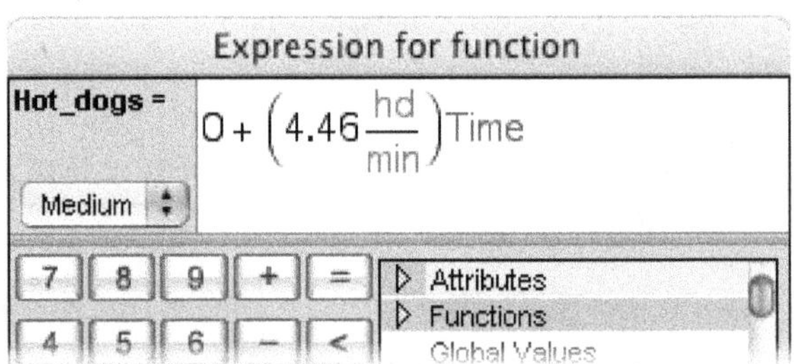

Q3 What equation does the line have?

Q4 How does the equation you found in Q3 relate to the recursive formula you entered in step 3 to generate the data points?

6. Drag the movable line out of your way and click on the true line. Drag a red dot along this line. The dot's coordinates are displayed next to it and in the status bar.

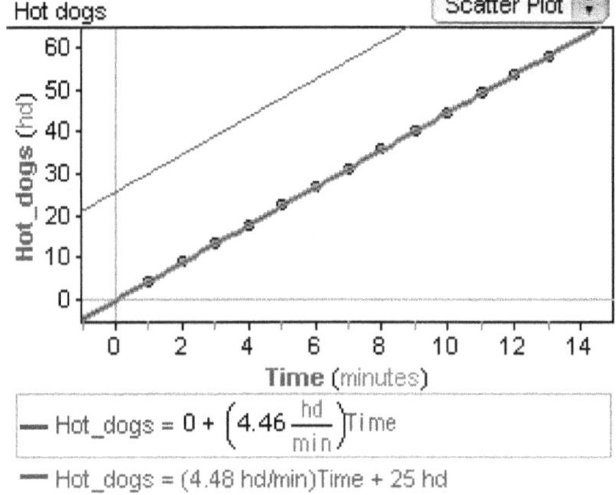

Q5 About how many hot dogs had Takeru eaten after 3.5 minutes?

Use the up-and-down arrow that appears near the center of the line to move it without changing the slope.

7. You decide to compete with Takeru. Because you can eat at a speed of only 2.3 hot dogs per minute, you're offered some hot dogs as a head start. Adjust the movable line to reflect your eating rate of 2.3 hot dogs per minute. Then adjust the line, without changing the slope, so you tie Takeru at 12 minutes.

Q6 What is the equation of your movable line?

Q7 How many hot dogs must you be given as a head start at time 0 in order to tie Takeru?

Q8 How does the equation represent your head start and eating speed?

Q9 Why is this equation said to be in *intercept form?*

EXPLORE MORE

Make up some additional problems about the hot-dog–eating contest.

Intercept Form—Hot Dogs

Objective: Students will use movable lines on Fathom graphs to understand the slope of a line as a rate of change and as a coefficient in a linear equation. They will see the constant in a linear equation as the y-intercept of the corresponding line and learn about the intercept form of a linear equation.

Student Audience: Algebra 1

Activity Time: 30–40 minutes

Setting: Paired/Individual Activity

Mathematics Prerequisites: Students can read the coefficient of a variable from an equation, understand coordinates of points in the rectangular coordinate system, and understand slope of line as steepness.

Fathom Prerequisites: Students can open a new file; set the linear equation form in **Preferences;** set up attributes and add new cases to a case table; enter units and formulas, including recursive formulas, for attributes; and add a movable line to a scatter plot.

Fathom Skills: Students will use *caseIndex* in formulas, create a scatter plot, plot a function, and trace a function plot with a red dot.

Background: This information was found in March 2005 at sports.espn.go.com/espn/news/story?id=1834236.

Notes: As you check on working students, be alert to potential difficulties if they fail to realize that attributes and units are case sensitive. Ask questions that will help students distinguish between the variable *Hot_dogs*, which refers to the number of hot dogs eaten, and hd, the units. You might find it helps some students to lengthen the name of the attribute, or variable, to *Number_hot_dogs_eaten*. You might suggest that students make a table by hand before they make the Fathom table in step 2. Make sure students realize they can create new units and should click **Yes** to create a new unit named hd in step 2. If your students are not familiar with the recursive form of an equation and can't make meaning of the notation, suggest that in step 3 they enter a formula that has meaning for them, such as *Time* • 4.46.

As you visit groups and during discussion, give pairs opportunities to explain their thinking on Q4, Q7, Q8, and Q9. Students benefit from hearing others' explanations.

Q1 Answers will vary considerably.

Q2 $\frac{53.5}{12}$, or 4.46 hot dogs per minute

INVESTIGATE

Q3 Equations should be close to *Hot_dogs* = 0 hd + (4.46 hd/min) *Time.*

Q4 The coefficient of *time* is the number of hd/min in the recursive formula.

Q5 About 15.6 hot dogs

Q6 Equations should be close to *Hot_dogs* = 26 hd + (2.3 hd/min) *Time*

Q7 Answers will vary around the theoretical value of 26 hot dogs.

Q8 The constant is the line's y-intercept, the head start; the coefficient of time is the line's slope, the eating speed or the rate at which hot dogs are consumed.

Q9 The equation begins with the y-intercept. It then adds or subtracts the slope—amount of change in height (the vertical change) per unit—multiplied by the number of units.

Point-Slope Form—Men's High Jump

The world record for men's high jump has increased very consistently from 1971, until the record (as of 2005) was set in 1993. If the record height had continued to increase at the rate it did during those 22 years, what might the record be today?

INVESTIGATE

1. Open the Fathom document **High Jump Men.ftm.** You will see a table of men's records for competition high jumps since 1970.

Q1 For the data shown, what is the current record, when was it set, and who set it?

2. Drag down a new graph and make a scatter plot with *Year* on the horizontal axis and *Height* on the vertical axis.

The data appear linear, but finding the intercept form of a line through them could be difficult because the y-intercept is out of sight.

To open the formula editor, double-click on the collection to bring up the inspector, select the **Cases** panel, and click on the formula space next to *x*. Alternatively, select the table and choose **Table | Show Formulas,** then click on the formula space under *x*.

3. You can change the intercept by using the number of years before or after a specific year. A year near the center of the data, such as 1978, is a good choice. In the table, click on <*new*> and create a new attribute called x. Define a formula for this attribute: Year – 1978.

Nationality	Height	x
		Year – 1978

Formula for x

Year – 1978

4. Drag the attribute x to the horizontal axis of the graph, replacing *Year.*

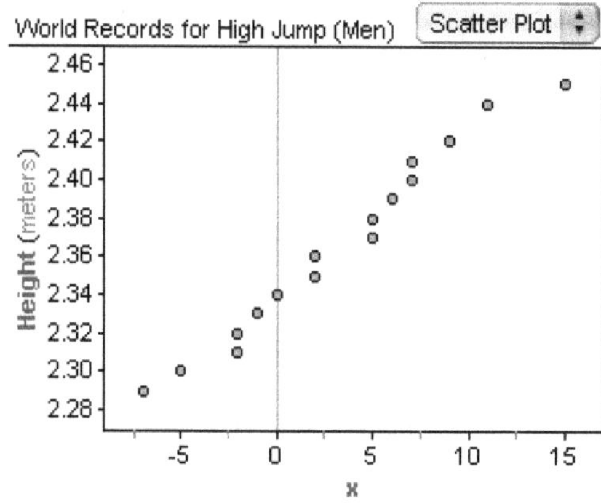

Q2 Where would the line of fit cross the vertical axis for these data?

5. Use a movable line to find a good line of fit that goes through the point (0, 2.34). Select the graph and choose **Graph | Plot Function,** then enter the same equation as the line you just found. (*Note:* You will need to enter units of m for both the intercept and the slope.)

6. Drag *Year* to the horizontal axis of the graph, replacing x. The numbers on the axis will change, and the line will vanish. In step 3, you created the attribute x using *Year* − 1978. To replace x with that expression, double-click on the function equation below the graph and replace the x in that equation with (Year – 1978).

Q3 The form $y = k + b(x - h)$ is called the *point-slope form* of a linear equation. Why do you think it has this name?

7. Notice where the coordinates of the point (1978, 2.34) appear in your equation. Select any data point that the line does not go through and note its coordinates. Edit the equation to replace the coordinates of the point (1978, 2.34) with the coordinates of the new point, but don't change the slope.

Q4 Explain what happened to the graph of the line.

Q5 Given the equation $y = 47 + 1.5(x - 28)$, name a point on this line without graphing or calculating anything.

Q6 If the trend established between 1971 and 1993 had continued, what would be the world record today?

The units are years, meters, and meters per year.

8. Add three sliders to your workspace; label them h, k, and slope; and include the correct units. Edit the function and replace the numbers by entering the slider variables Height = k + slope(Year – h). With the graph selected choose **Graph | Plot Value** and enter h. Then choose **Plot Function** and enter k. Begin by entering the values you used in your equation for *h, k,* and *slope.* Use the sliders to make adjustments in your model.

Q7 As you move the slider, how do the values of *h* and *k* change the graph?

Point-Slope Form—Men's High Jump

Objective: Students will use a movable line and new attributes to write the equation of a line in point-slope form.

Student Audience: Algebra 1

Activity Time: 20–30 minutes

Setting: Paired/Individual Activity (use **High Jump Men.ftm**) or Whole-Class Presentation (use **High Jump Men Present.ftm**)

Mathematics Prerequisites: Students can calculate the slope of a line and write equations in intercept form.

Fathom Prerequisites: Students can open Fathom files, add attributes with formulas, exchange attributes on scatter plots, plot functions and values, and work with sliders.

Fathom Skills: Students will learn to edit functions.

Notes: Traditionally, we teach the point-slope form by beginning with the equation for slope and using a generic point (x, y). This activity uses the replacement of attributes to create the formula. This approach may appeal to some students and will provide another way of thinking for all students. It is valuable preparation for transformations.

As you visit groups, be alert to possible student errors. Students might forget that attributes and units are case sensitive, and they need to check the spelling of words such as *minutes* to make sure Fathom does not create new units. There are colors in Fathom to alert users to correct spellings. Ask students for the clues they see or tools they find to help them check their typing.

For a Presentation: If your class works through the activity together, ask a student to run the presentation computer. Before x is dropped on the graph in step 4, ask Q2. Encourage discussion of Q4 and Q6. You might use the document **Men's High Jump Present.ftm,** which has the sliders. Scroll to the right in the document to find the sliders; inspectors have been used to set reasonable boundaries. Ask for several student explanations for Q7.

INVESTIGATE

Q1 Javier Sotomayer set the record in 1993, at 2.45 m.

Q2 Answers will vary; 2.34 m is a good choice.

Q3 The point-slope form of the equation is built from the coordinates of one point and the slope.

Q4 The line jumped up or down to intersect the y-axis at a different point but remained parallel to the first line.

Q5 The point that the equation is built from is (28, 47).

Q6 About 2.6 m in 2006

Q7 Changing (sliding) h seems to move the line left and right. Changing (sliding) k seems to move the line up and down. If students have trouble getting the line controlled by the sliders to appear, encourage them to open the slider inspector (click on the thumb) and set the upper and lower limits for the sliders; h should have the limits of the years in the data, k can be limited to between 0 and 3, and the slope will be between 0 and 0.01.

EXTENSION

Ask students to find out whether the record set in 1993 has been broken yet. If it has, add that point to the scatter plot. How does that change student estimates for the line of fit?

Point-Slope Form—Life Expectancy

Life expectancies are the average (mean) lengths of life for persons in various groups. For example, the life expectancy of a person born in 1940 is 62.9 years. This means that the anticipated average age of death of all people born that year is 62.9 years (though many of them haven't died yet). Life expectancy is predicted on the basis of many factors. Considering past data on life expectancies, what would you predict the life expectancy of someone born in 2010 to be?

Q1 Make a conjecture about the life expectancy of someone born in 2010.

Q2 Do you think the life expectancy of females born that year will be more or less than that of males born that year?

INVESTIGATE

1. Open the Fathom document **Life.ftm** and drag a graph from the shelf. Drag the *Birth_Year* attribute to the horizontal axis and one of the other attributes (*Female, Male,* or *Combined*) to the vertical axis to make a scatter plot.

To make a prediction about 2010, you want a line that seems to fit the data set. It needs to be plotted, rather than movable, so you can slide a red dot along it. However, because the data points aren't in a straight line, you'll want to be able to manipulate the line easily. For that, you'll use a slider.

To find the slope, divide the difference between the life expectancies $(y_1 - y_2)$ by the difference between the birth years $(x_1 - x_2)$.

Q3 Write down the coordinates of one of the data points. Select a second point and use it to calculate the slope of the line between the points.

Q4 How many birth years are between the points? How many birth years are between your first point and another point with horizontal coordinate *Birth_Year?* (This answer will be an expression that includes *Birth_Year* in place of a number.)

Q5 How many life expectancy years are between your two selected points? How can you use the slope (Q3) and the number of birth years between the points (Q4) to calculate this number?

Q6 Find an expression for the rise or fall (change in life expectancy) between your start point and a point with horizontal coordinate *Birth_Year;* use the variable *Slope* in place of the number for slope. (This answer will be an expression using both *Birth_Year* and *Slope* in place of the numbers you used in Q5.)

Q7 How can you find the vertical coordinate of your second point using the vertical coordinate of your first point and the answer to Q5?

This answer is an expression using both *Birth_Year* and *Slope* in place of the numbers you used in Q7.

Q8 How can you find the vertical coordinate of any point on the line using the vertical coordinate of your first point and the answer to Q6?

Double-click the slider's thumb to open the inspector.

2. Drag a slider from the shelf. The slider name, *V1*, will be selected; rename the slider by entering Slope. Adjust the scale on the slider or set *Upper_* and *Lower_* limits in the slider's inspector so that it runs from about 0 to about 0.4.

3. Choose **Plot Function** from the **Graph** menu. Enter the expression you wrote for Q8, using the variable *Slope*. The slider will now control the slope.

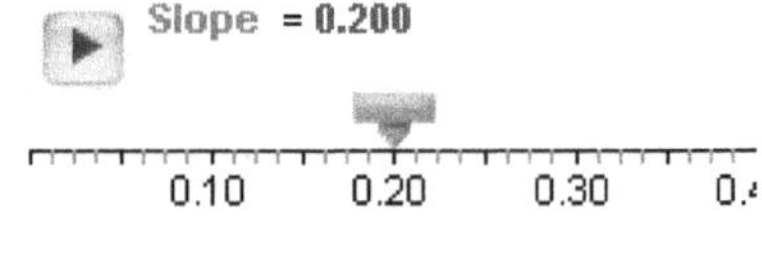

You might need to adjust the scale on the graph.

4. Adjust the slider so that the line appears to fit the data points as closely as possible.

5. If you think a line through a different data point will represent the data better, double-click on the line's equation and enter coordinates of that other data point. Adjust the line this way until you think you can make a good prediction.

Q9 Record the equation. This graph shows one possible equation.

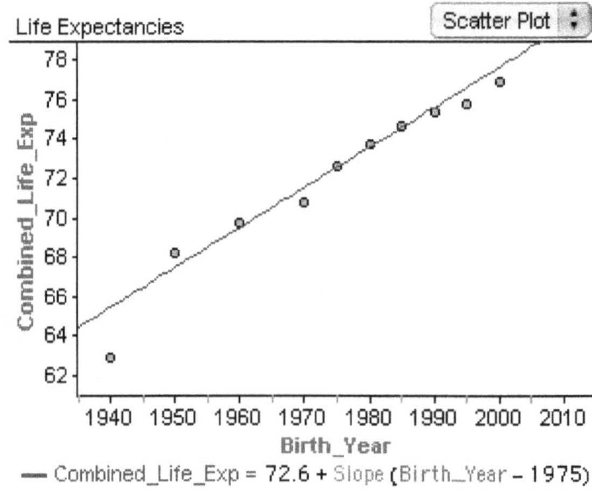

6. Click on the line and move the red dot, watching the coordinates in the status bar.

You might need to rescale the vertical axis.

Q10 What prediction does your line make for life expectancy in 2010?

Q11 Why might the form you've been using, $y = y_1 + b(x - x_1)$, be called the *point-slope form* of a linear equation?

EXPLORE MORE

1. To adjust your line even more flexibly, you can use sliders for the point to pass through, rather than being limited to a data point. You might use sliders called *x1* and *y1* for the two coordinates of the point. Change your formula accordingly. What equation do you think gives the best prediction, and what is that prediction?

2. Adapt your graph, or make a new one, with both *Male* and *Female* on the vertical axis. Do so by having one of the two attributes already on this axis and then dragging one toward the axis. A plus sign appears; drop the second attribute on the plus sign. Both data sets will appear on the scatter plot. Find an equation to predict whether the life expectancy for females will still be greater than that for males and how much the life expectancies will differ. Are the life expectancies getting farther apart or closer together? What measure(s) indicate(s) that?

Objective: Students develop and use the point-slope form of the equation of the line.

Student Audience: Algebra 1

Activity Time: 35–50 minutes

Setting: Paired/Individual Activity, Exploration, or Whole-Class Presentation (use **Life.ftm** for any setting)

Mathematics Prerequisites: Students can calculate slope from two points.

Fathom Prerequisites: Students can open an existing Fathom file, make a scatter plot, plot a function, and edit the equation of a plotted function.

Fathom Skills: Students will use a slider and trace a function graph with a red dot.

Notes: This activity introduces and explains the point-slope form as $y = y_1 + b(x - x_1)$, where y_1 represents some starting value for the dependent variable, b is the rate of change, and $x - x_1$ is the distance moved from the starting point. The location of the points on the line are given in terms of some starting point (x_1, y_1) and the movement from that point.

If students are stuck on how to write an expression, have them think about what they did with specific values before they try to generalize. Suggest they repeat Q3 and Q5 for more than one choice of second point before introducing a variable in the expression in Q4 and Q6. As you listen in on groups that are struggling with Q7, you might suggest that they substitute the points they chose in the question.

As groups share their predictions, ask them to describe the points and equations that lead to those predictions and explain why they chose those points.

Exploration: For students who don't need step-by-step instructions, you might use the Exploration on page 80. Students using the Exploration may or may not use the point-slope form.

For a Presentation: As your class discusses the answers, use several points for Q3 and Q5 before writing the expressions with the variables. Once you have a line of fit drawn, ask students to suggest different points the line could go through to give a better line of fit. For each of those points, write an equation and predict the 2010 life expectancy.

Q1 Answers will vary considerably.

Q2 Women will probably have a longer life expectancy.

INVESTIGATE

Q3 Students might choose any data point. The next few answers are based on the choice of (1970, 70.8) and (1990, 75.4) for (*Birth_Year, Combined*). The line between the points has slope $\frac{75.4 - 70.8}{1990 - 1970}$, or 0.23.

Q4 1990 − 1970, or 20, years between the points; *Birth_Year* − 1970 or 1970 − *Birth_Year* (The form *Birth_Year* − 1970, which gives positive values after 1970 and negative values before 1970, is more conventional and will be used below.)

Q5 The difference in life expectancy rose 75.4 − 70.8, or 4.6, but also 0.23(20) = 4.6; multiply the slope times the number of years between the two points.

Q6 *Slope* (*Birth_Year* − 1970)

Q7 70.8 + 4.6 = 75.4; add the vertical coordinate to the answer to Q5.

Q8 70.8 + *Slope* (*Birth_Year* − 1970)

Q9 One good line has the equation *Combined* = 72.6 + 0.17(*Birth_Year* − 1975).

Q10 The line in Q9 predicts a life expectancy of about 78.6 in 2010. Men's life expectancy will be about 76 years, and women's will be about 83 years.

Q11 This form incorporates the coordinates of a point (x_1, y_1) as well as the slope (b).

EXPLORE MORE

1. Answers will vary a good deal. Some will be close to the answer to Q10.
2. Most models will show the male life expectancy to have a smaller slope than the higher female life expectancy, so they would be moving apart.

Life expectancies are the average (mean) lengths of life for persons in various groups. For example, the life expectancy of a person born in 1940 is 62.9 years. This means that the anticipated average age of death of all people born that year is 62.9 years (though many of them haven't died yet). Life expectancy is predicted on the basis of many factors.

MAKE A PREDICTION

Consider data in the Fathom file **Life.ftm** and predict the life expectancy of someone born in 2010. You can decide whether to predict for *Females, Males,* or *Combined.*

INVESTIGATE

For your prediction, you can use a line that is a good fit for the data points and find the y-coordinate for the point (2010, y). To plot a line along which you can drag a red dot, choose a point through which the line will pass and think about how much the line will rise or fall from that point by the time given by *Birth_Year.* Sliders will be very helpful in adjusting your plotted line.

EXPLORE MORE

With one of the two attributes already on this axis, drag the other one toward the axis. A plus sign appears; drop the second attribute on the plus sign. Both data sets will appear on the scatter plot.

Adapt your graph, or make a new one, with both *Male* and *Female* on the vertical axis. Find an equation to predict whether in 2010 the life expectancy for females will still be greater than that for males and how much the life expectancies will differ. Are the life expectancies getting farther apart or closer together? What measure(s) indicate(s) that?

Linear Modeling—Dissolved Oxygen

All animals, including fish, require oxygen to breathe. Temperature, water quality, and other factors can affect the amount of dissolved oxygen in water.

Q1 As the temperature of water increases, do you think the water will hold more oxygen or less oxygen? Why?

To make a prediction about oxygen levels in water from some data that lie in a somewhat straight line, you look for a line of fit. Several lines fit the data fairly well. You will develop a method anyone can follow to fit a line to approximately linear data.

INVESTIGATE

1. Open the Fathom document **Oxygen.ftm,** which gives dissolved oxygen (*DO*) at various temperatures from a sample of lake water. Drag a graph from the shelf.

Your first goal is to plot a line that will allow you to predict a temperature at which *DO* is 0 parts per million (ppm).

With the graph selected, choose **Graph | Add Movable Line.**

2. Make a scatter plot of the data (*Temperature* on the horizontal axis, *DO* on the vertical axis), and add a movable line. Adjust the movable line until it fits the points fairly well. Adjust the scales on the axes so you can see where the line crosses the horizontal axis.

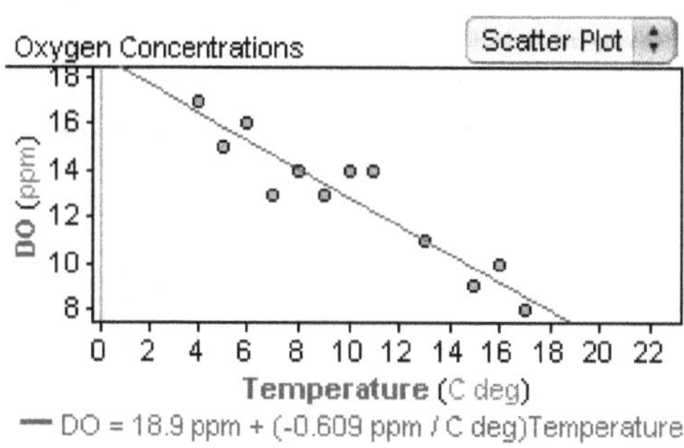

The equation for your movable line is displayed below the scatter plot.

Q2 What is the equation of the movable line?

Q3 At what value for *Temperature* do you predict *DO* will be 0 ppm? Compare your answer with those of your classmates.

Your answer may be similar to those of others, but probably not all answers are the same. Your goal is to find a method of getting the line by using specific steps, not by eyeballing, so that others can replicate your process. Here's one approach that makes the line go through the middle of the data.

3. From the **Graph** menu, choose **Plot Value,** and enter median(temperature). The result will be a vertical line going through the median of the temperatures on the horizontal axis.

4. Choose **Plot Function** from the **Graph** menu and enter median(DO).

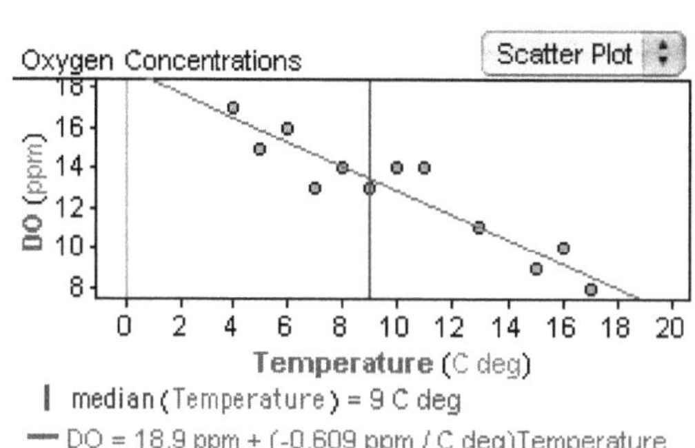

You'll see a horizontal line through the median of the values. Adjust your movable line to pass through the intersection of these median lines and adjust the slope as needed to fit the data points.

Q4 What is the equation of the movable line now?

Statistical
One Attribute
count
first
iqr
last
max
mean
median
min
percentile
popStdDev
popVariance
product
proportion
Q1
Q3

Q5 Now what value of *DO* do you predict when the temperature is 0°C? Do your classmates have a similar value?

You still had to eyeball the slope. A method that would completely determine the line would use two points instead of just the one point determined by the medians.

5. Remove the median lines by double-clicking on their formulas and deleting them in the formula editor. Use other statistics from *Temperature* and *DO* to plot values and functions to get two points through which the line will pass. To see other possible statistics, look in the formula editor under **Functions | Statistical | One Attribute.**

To return to the original data, you can select the collection and then choose **File | Revert Collection.**

6. Drag some data points around to change the collection. As needed, adjust your methods to get lines that fit a variety of data. Then return to the original data.

Q6 Describe what you did, the resulting equation, and the prediction it led to.

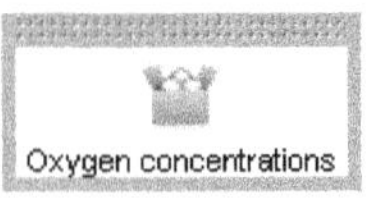

The DO concentration is often used to check water quality. The second collection, South Hood Canal, contains average winter measures of dissolved oxygen as the quality of the South Hood Canal in Olympia, Washington, continues to worsen.

You may need to scroll down to find the second collection.

7. Create a new graph and place South Hood Canal's *Year* on the horizontal axis and *Oxygen* on the vertical axis.
8. Test and refine your method of defining a line of fit on this new data.

Q7 Compare your line with your classmates' lines generated by different methods.

Q8 Develop a method for finding a standard line of fit from the first quartile (Q1 in the statistical functions that Fathom calculates) and the third quartile (Q3).

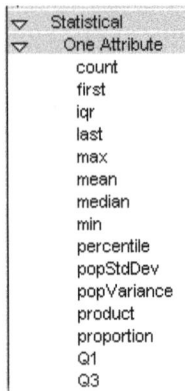

When you select Q1 in the formula editor, the **Help** panel describes the first quartile statistic.

EXPLORE MORE

Make scatter plots and test your methods on the data in **Life.ftm, High Jump Women.ftm,** or **High Jump Men.ftm.**

Linear Modeling—Dissolved Oxygen

Objective: Students will see the reason for developing a standard method for finding lines of fit, and they will use quartile points to find a line of fit.

Student Audience: Algebra 1

Activity Time: 25–35 minutes

Setting: Paired/Individual Activity (use **Oxygen.ftm**)

Mathematics Prerequisites: Students understand how equations relate to linear graphs and the x-intercept of lines.

Fathom Prerequisites: Students can open a Fathom document, make a scatter plot, and create and adjust a movable line.

Fathom Skills: Students will learn how to adjust scales of axes and to plot values and functions.

Notes: If your students can solve linear equations, you might ask them to give x-intercepts that are more exact than the graphical approximations. When most students have answers to Q2 and Q3, you might ask them to write the equations and predictions on the board for comparison. Do the same with Q4 and Q5. Answers to Q6 and Q7 can be compared during sharing once the investigation is complete. During discussion, help students see that whenever a prediction is made from approximately linear data, there will be variation. If the predictions from several data sets are to be compared, it's necessary to have a standard method of finding the line of fit. During sharing, ask students to explain their standardized methods. You might choose two methods and use them and the Q-line on the data suggested in Explore More. Compare the results of the three methods.

Q1 Answers will vary. As the temperature increases, water will hold less oxygen. Heat causes molecules to move faster, so more of the oxygen will escape to the air.

INVESTIGATE

Q2 Answers will vary considerably. One equation is $DO = 18.7$ ppm + (–0.582 ppm/Celsius) *Temperature.* Another is $DO = 18.9$ ppm + (−0.609 ppm/Celsius) *Temperature.*

Q3 Answers will depend on the equation in Q2. The equations above yield temperatures of about 32.1°C and 31.0°C, respectively.

Q4 Answers may vary widely. One line with almost the same number of data points on each side and passing through the intersection of medians at (9, 14) has equation $DO = 20$ ppm + (−0.902 ppm/Celsius) *Temperature.*

Q5 The equation in Q4 yields a temperature of about 22.2°C.

5. Students might use quartiles, the median or mean and one quartile, or maximums or minimums; or they might use the means of the lower half and the upper half of each data set. If students are having trouble understanding what to try, point out that the explanations of the statistics in the formula editor are displayed in the window when you put the cursor on a statistic.

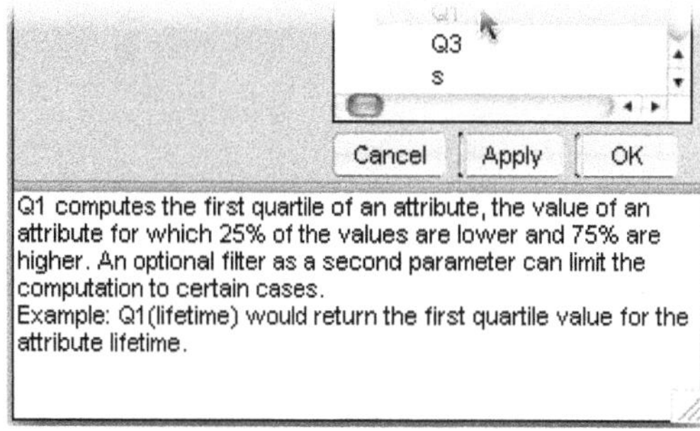

Q6 Answers will vary widely. For students who are frustrated with Q6, suggest they complete Q8 first. One possibility is to use quartiles. The intersections of lines through quartiles are at (6, 15) and (13, 11). The line through these points has equation $DO = 18.4 - 0.57$ *Temperature.* This equation has $DO = 0$ when *Temperature* is about 32.2°C. Another method might use the maximum and minimum points and yield $DO = 19.8 - 0.695$ *Temperature* (where $DO = 0$ when *Temperature* is about 28.8°C). However, outliers greatly affect maximum or minimum values, so the method of using max( ) and min( ) can misrepresent a data set with outliers. Because quartiles aren't strongly affected by outliers, they provide a better method.

Q7 Answers will vary, depending on methods. For this data set the method of max( ) and min( ) gives a line, *Oxygen* = 5.55 − 0.25(*Year* − 1998), above most of the points. The line generated by the method using quartiles, called the *Q-line*, is
Oxygen = 4.99 − 0.2325(*Year* − 1999)
It fits the points well.

Q8 Use the line through the points (x_{Q1}, y_{Q1}) and (x_{Q3}, y_{Q3}) to graph the standard Q-line. See Q7 for the equation.

EXPLORE MORE

The Q-lines for the other data sets are Life Expectancy:
Female = 73.1 + 0.19(*Birth_year* − 1960)
Male = 66.6 + 0.173(*Birth_year* − 1960)
Combined = 69.7 + 0.19(*Birth_year* − 1960)
Women's High Jump:
Height = 1.96 + 0.01125(*Year* − 1976)
Men's High Jump:
Height = 2.325 + 0.00941(*Year* – 1976.5)

Equation Solving—Cryptography

Cryptography is the mathematics of coding and decoding. To code a message, there is a "key" that changes the text in some way. In this activity, you will change each letter to a number from 1 to 26. Your key will change that number to another number. You can then send your message of numbers to a friend. Your friend will decode the message by reversing the process with a "reverse key."

INVESTIGATE

Click on the slider's thumb to open its inspector.

1. Open a new Fathom document and drag a slider from the shelf for the *Start* number and another slider for the *Encoded* number. In the slider inspectors, restrict them to multiples of 1 and set *Lower_* and *Upper_* to keep the *Start* values between 1 and 26.
2. For now, give the *Encoded* slider the formula 7 • Start + 5. This is the code's key.

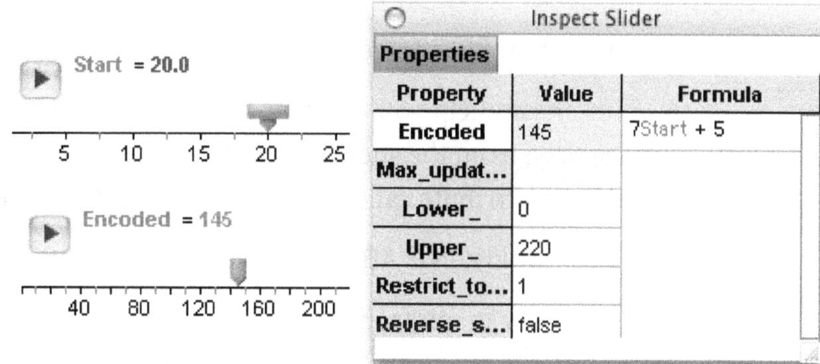

3. Make up a short message. Then translate each letter into a number, using 1 for A, 2 for B, and so on, up to 26 for Z.

Q1 What is your message, and what is its translation into code?

Q2 As you put the *Start* slider on each number of your message in turn, what numbers are encoded? This is the code you send to your friend.

Restrict the *Decoded* values to multiples of 1.

4. Now see how your friend might set up a slider to decode. Drag a slider from the shelf and enter the name Decoded. Think about what the encoding process did to the *Start* number, then give the *Decoded* slider a formula to undo this process. Be sure the *Decoded* slider moves to the same values as the *Start* slider.

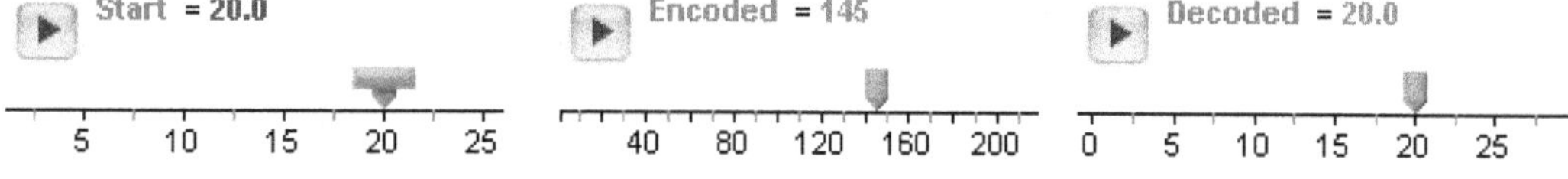

Q3 In words, what does the encoding process do to the *Start* number?

Q4 What reverse key formula did you give to the *Decoded* slider to undo the encoding process?

Encoding a single number *Start* is equivalent to evaluating the expression $5 + 7(Start)$ at that number. Decoding a single encoded number is equivalent to solving the equation $5 + 7(Start) = Encoded$ for the number *Start*. For example, if the *Encoded* number is 68, then your friend solves the equation $5 + 7(Start) = 68$ for the value *Start*, getting $Start = 9$ (the letter I).

5. Change the key in the *Encoded* slider to one of your own choosing. Change the reverse key to undo it.

Q5 What key and reverse key did you use?

Q6 Encoding with your key involves evaluating what expression?

Q7 Decoding with your reverse key involves solving what equation?

The formula can be entered into the case table (with the table selected, choose **Table | Show Formulas**) or on the **Cases** panel of the collection's inspector. (Show the inspector by clicking on the collection's white space, not its name.)

6. Fathom can help you translate letters into numbers to be encoded. Drag a new case table from the shelf, choose **Collection | New Cases,** and add one case. Make an attribute called In_Letter. Give it the formula

midstring("ABCDEFGHIJKLMNOPQRSTUVWXYZ", Start,1)

7. Drag a new graph from the shelf and drag *In_Letter* to the horizontal axis. As you drag the *Start* slider, the corresponding letter will appear below the bar in the graph.

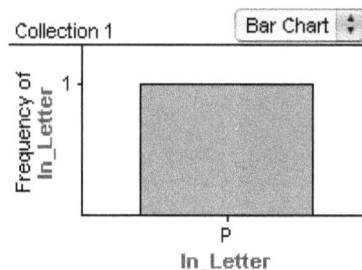

Q8 What is the formula for *In_Letter* doing?

Q9 Make up another short message. Use the bar graph to encode it. What are your message, your key, and the encoded message?

8. You can check your friend's decoder by adding an *Out_Letter* attribute to the case table, giving it an appropriate formula, and making a bar graph for it.

Q10 What formula did you use for *Out_Letter?*

EXPLORE MORE

1. Extend your coding to include more than the 26 capital letters. You might use punctuation marks, a space, lowercase letters, or another symbol. Report on a message you write, your key, and the encoded message.
2. Explore how to use Fathom's modulo function to translate an encoded number into a modular number, then into a letter shown on a bar graph.

Equation Solving—Cryptography

Objective: Students will practice undoing a linear expression to solve an equation and will see how solving an equation is the reverse of evaluating an expression.

Student Audience: Pre-algebra, Algebra 1

Activity Time: 30–40 minutes

Setting: Paired/Individual Activity

Mathematics Prerequisites: Students understand the order of operations and can solve simple linear equations.

Fathom Prerequisites: Students can use sliders.

Fathom Skills: Students will edit attribute formulas and create a slider formula.

Notes: Students who are intrigued by codes will enjoy this activity, which uses sliders as encoding and decoding machines. As you visit groups, find opportunities to ask about the connection between order of operations and the way students determined the reverse key. After student pairs investigate, ask them to share their answers to Q5–Q10.

This activity intentionally has students send numbers as codes rather than translate the numbers back into letters by "wrapping," or subtracting multiples of 26. Wrapping requires modular arithmetic and is therefore more complicated to undo. Explore More 2 includes a suggestion for students to explore the more complicated ideas.

INVESTIGATE

Q1 Messages and their translations into numbers will vary.

Q2 Encoded messages will vary.

Q3 The starting number is multiplied by 7, then 5 is added. Emphasize that 5 is not added to 7 first—in the order of operations, multiplication precedes addition.

Q4 A reverse key will undo in reverse order: Subtract 5 and then divide by 7. So, the formula is *Decoded* = $\frac{Encoded - 5}{7}$.

Q5 Answers will depend on student keys.

Q6 Encoding is evaluating the expression in the student's key.

Q7 Decoding is solving for *Start* in the equation [student key] = *Encoded.*

From this point forward, answers will depend on each student's expressions. Sample answers are given to indicate the type of answer.

Q8 The formula is finding the string of letters ("midstring") inside the alphabet that begins at *Start* and goes for one letter. That is, it produces a single letter that corresponds to the *Start* number.

Q9 Answers will vary.

Q10 midstring("ABCDEFGHIJKLMNOPQRSTUVWXYZ", *Decode*, 1)

EXPLORE MORE

1. The encoder will take on values beyond 26, depending on how many symbols are appended. The string of characters in the midstring commands will give all the symbols.
2. The numbers will depend on the symbol set chosen. With the original 26 letters, coded numbers more than 26 will be changed to a number from 1 to 26, using a Fathom expression such as modulo(*Encoded,* 26). For example, add a new attribute to the collection called *Encoded_Letter* and give it this formula: midstring("ABCDEFGHIJKLMNOPQRSTUVWXYZ", modulo(*Encoded*, 26), 1).

3

Solving Inequalities and Systems

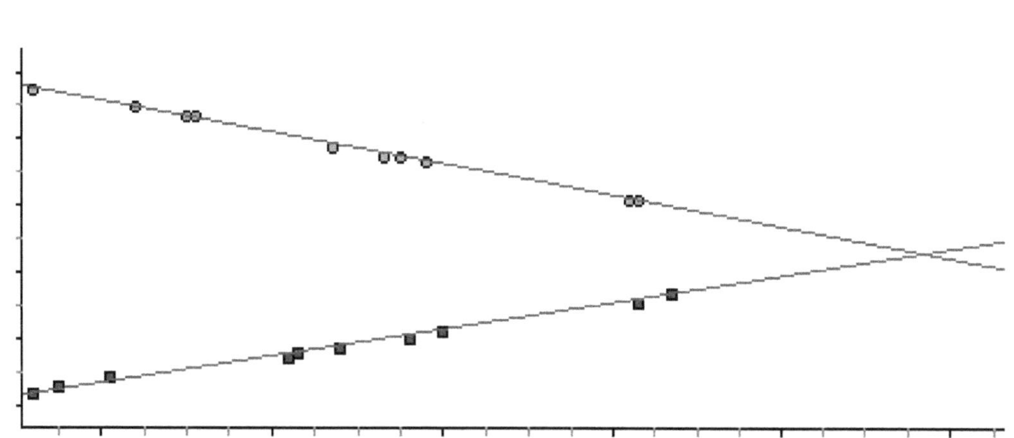

Inequalities—Strange Diet

As Roy leaves wrestling practice, he receives these strange instructions from the new coach: "Your next meal should have more than 80 grams of fat, at least 1500 calories, less than 3090 milligrams of sodium, and no more than 18 grams of sugar."

Roy is off to Fast Burger to get in on the special price for three of the same sandwich and a large order of fries. Your job is to help Roy decide which sandwich might meet these instructions.

Q1 An *inequality* is a statement, like that of the coach, that describes how one quantity is different from another, such as "more than," "at least," or "no more than." Name at least three inequality instructions you have given or received recently.

INVESTIGATE

1. Open the Fathom document **Diet.ftm.** You'll see a case table and, below the table, a graph, showing a dot plot of the fat content of foods. Move the cursor over the points in the dot plot and read the information in the status bar.

Deluxe Burger (29 g)

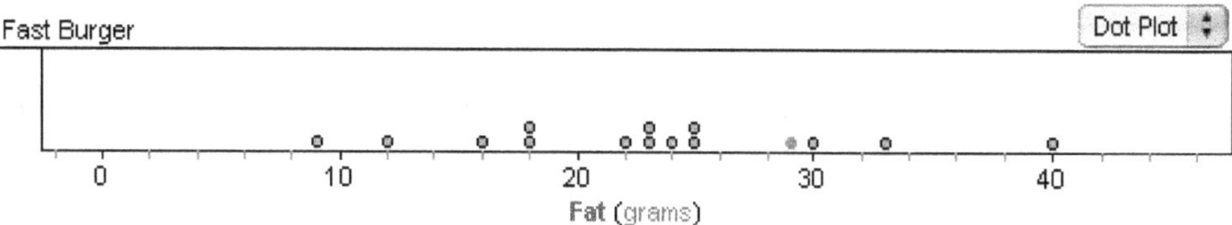

Q2 Which sandwich has the most fat? Which has the least fat?

You know that Roy will order three of the same sandwich and a large order of fries. If you check the last entry in the table, you'll find that the fries have 25 g of fat.

Inequalities can be represented with an inequality sign ($<$, $\leq$, $>$, or $\geq$), just as an equation is a statement made with an equal sign. The coach's first instruction can be written as an inequality: *Fat* is a variable representing the fat in one sandwich; so the fat in three sandwiches plus the fat in the fries (25 g) must be greater than 80 g: $3Fat + 25\text{ g} > 80\text{ g}$.

Be sure to enter the multiplication sign; it will not always show up in the formula editor.

2. To see only the data points that satisfy this condition, you can make a filter in Fathom. Click on the graph to select it. Choose **Object | Add Filter.** This kind of filter lets through points meeting the condition you enter; so enter 3 • Fat + 25g > 80g.

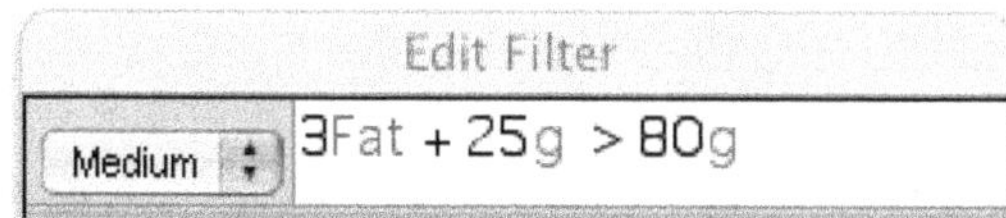

When you click **OK,** some of the points disappear from the graph.

Q3 Which sandwich among those remaining has the least fat?

You solve an inequality as you do an equation, except you reverse the direction of the inequality when you multiply or divide by a negative number.

To enter a symbol for greater than or equal to, hold down the Option key (Mac) or Ctrl key (Win) while entering the greater than symbol.

Q4 Solve the inequality $3Fat + 25\text{ g} > 80\text{ g}$ for *Fat*. How does the solution relate to the points that remain on the dot plot?

3. Make a duplicate of the graph by selecting it and choosing **Object | Duplicate Graph.** On the new graph, drag the attribute *Calories* to replace *Fat*. Double-click the filter line and edit it to match the second instruction—at least 1500 calories. (If all points disappear, you may have omitted necessary units.)

Q5 Solve the inequality to find the restrictions on calories.

Q6 Which remaining choices have the fewest calories? Which have the most calories?

Q7 Repeat step 3 for the remaining two instructions from the coach, and answer Q5 and Q6 for the attributes *Sodium* and *Sugars*.

Q8 Does the graph of *Sodium* include the point for Spicy Crisp Chicken Burger (920 mg)? Why or why not?

Q9 Does the graph of *Sugars* include the point for the Deluxe Crisp Chicken Burger (6 g)? Explain.

Q10 In turn, click on each point in the dot plot for *Sugars* and see if it is highlighted in the other graphs. Which sandwiches meet all the conditions the coach gave Roy?

4. Select the case table and choose **Add a Filter.** Using parentheses, enter each of the four solved statements, separated by *and*. (Remember the units.)

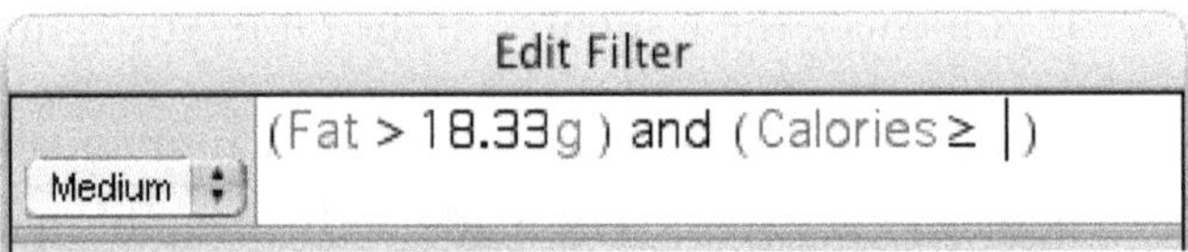

Q11 Does the table information agree with your answer to Q10? What is your recommendation to Roy?

EXPLORE MORE

Select one of the sandwiches in the collection. (Remove the filter to see all the cases.) Describe that sandwich using four or more inequalities about its attributes. Exchange inequality information, but not the name of the sandwich, with a classmate. Determine which sandwich your classmate has described.

Inequalities—Strange Diet

Objective: Students will explore solving inequalities for one variable by first looking at solutions to inequalities on a number line. They will translate different types of inequality statements from context to symbols, solve the inequalities, then interpret the solution in the context of the problem.

Student Audience: Algebra 1

Activity Time: 50–60 minutes

Setting: Paired/Individual Activity (use **Diet.ftm**)

Mathematics Prerequisites: Students can solve equations.

Fathom Prerequisites: Students can open a Fathom document and read a dot plot.

Fathom Skills: Students will learn how to duplicate graphs and add and edit filters for graphs and case tables.

Notes: The humorous context does not represent real nutritional limits. You might generate some discussion about what considerations would be reasonable when ordering a meal. Number lines are represented by dot plots, with inequalities entered as filters. As dots disappear and reappear, students see the difference between strict and weak inequalities. Pair color-blind students with students who can easily see the red. Be alert to potential difficulties when students fail to use the correct units. Student sharing can center around explanations for Q8–Q11. Interesting Explore More inequalities and the process of determining the sandwich can also be shared.

Q1 Allow some time for students to reflect that they are surrounded by decisions that must be made based on inequalities. Inequalities can guide us, whether they give pages in an English assignment or words in a newspaper advertisement. Some students may note that Q1 is one example.

INVESTIGATE

Q2 The Double Big Burger has the most fat (40 g), and the Regular Burger has the least (9 g).

Q3 The Crisp Chicken Burger, with 22 g of fat

Q4 $Fat > \frac{55}{3} \approx 18.33$ g; all the points on the graph are to the right of the solution.

Q5 $Calories \geq \frac{980}{3} \approx 326.67$ Cal

Q6 Both the Fish Burger and the Grilled Chicken Burger, with 400 Cal, have the fewest. The Double Big Burger, with 730 Cal, has the most.

Q7 $Sodium < 920$ mg; $Sugars \leq 6$ g. The item lowest in sodium is the Large French Fries, with 330 mg of sodium; the burger lowest in sodium is the Regular Burger, with 530 mg. The highest is the Crisp Chicken Burger, with 760 mg. The item with the lowest sugar is the Large French Fries, with 0 mg of sugar; the burgers with the lowest sugar are the Crisp Chicken Burger and the Spicy Crisp Chicken Burger, both with 5 g. The highest amount of sugar is in the Deluxe Crisp Chicken Burger, with 6 g.

Q8 No, it is the border point and is not included in strictly less than.

Q9 Yes, it is the border point and is included in less than or equal to.

Q10 The only sandwich choice that fits all the requirements is the Crisp Chicken Burger.

4.

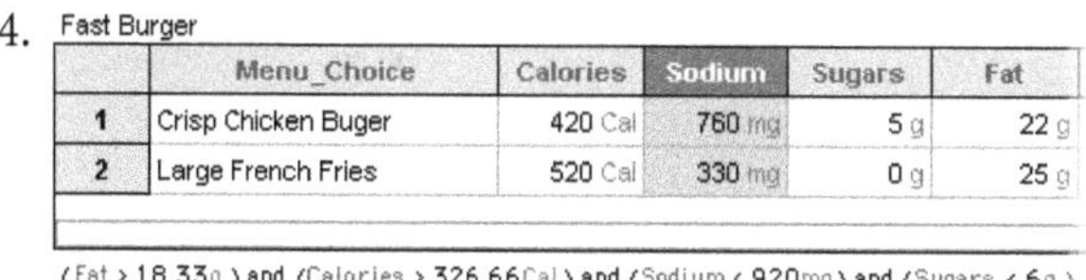

Fast Burger

	Menu_Choice	Calories	Sodium	Sugars	Fat
1	Crisp Chicken Buger	420 Cal	760 mg	5 g	22 g
2	Large French Fries	520 Cal	330 mg	0 g	25 g

(Fat > 18.33g) and (Calories ≥ 326.66Cal) and (Sodium < 920mg) and (Sugars ≤ 6g)

Q11 Yes; the table should show only the one sandwich. Students may suggest that four orders of fries would also fit the conditions.

Absolute Value—Radio Contact

While tracking communications from smugglers in a jet, federal agents recorded a conversation. The agents found that at 3 minutes past the hour, the plane was directly over a particular tower. The agents tell you that due to the range of the radio in that tower, any conversation between the plane and the tower must have taken place within 5 minutes on each side of the time the plane was directly overhead. Your goal is to see how an inequality can be solved to find the times at which this conversation could have taken place.

Q1 Relative to the hour, when could the conversation have taken place?

INVESTIGATE

To express the times within 5 minutes of the number 3, you can use an absolute value and an inequality. The symbol $|x - 3|$ refers to the amount of time between x and 3. So solutions to the inequality $|x - 3| \leq 5$ are those times within 5 minutes of 3.

1. Open the Fathom document **Contact.ftm.** You will find a dot plot that looks like a number line.

In Fathom, $|x|$ is written abs(x).

To enter $\leq$, hold down Ctrl (Win) or Option (Mac) while pressing <.

2. Select the Minutes dot plot and choose **Add Filter** from the **Object** menu. Type in the inequality as a filter.

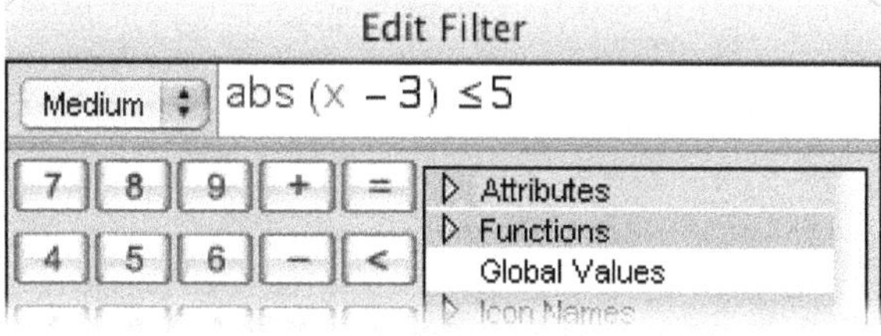

Q2 Describe, in words, the numbers that are solutions to the inequality. Do they agree with your answer to Q1?

With the graph selected, choose **Object | Duplicate Graph.**

Q3 Write your description, using two inequalities joined by the word *and* or *or.*

3. Make a duplicate of the graph and move it directly below the first.

Double-click on the current filter to open it in the formula editor.

4. On the second graph, change the filter to your statement in Q3. Be sure the two filters give the same solutions.

Q4 What instructions for the steps of solving an inequality would you give to somebody who is trying to change the filter $|x - 3| \leq 5$ to a filter that uses the word *and* or *or*?

Q5 Change the first filter, still using an absolute value inequality, to include the points—and only those points—not currently included in the graph. What inequality did you use?

Q6 How would you change the second filter, not using an absolute value, to give the same result as the new first filter?

Q7 What steps would you recommend for solving the inequality in the first filter to change it to the two statements in the second?

Sometimes a situation is modeled by an absolute-value inequality that is more complex than the one described above. The absolute value of a difference, though, still represents the space between the things being subtracted. For example, suppose you want to model the pressure at which an alarm is activated. You know that the alarm will sound when the pressure, represented by x, is 25 psi (pounds per square inch) or more away from the recommended pressure multiplied by a scale factor, which is determined by the temperature.

Q8 Suppose the recommended pressure is 300 psi and the space between the pressure and the recommended pressure must be scaled by $\frac{5}{6}$. What's an inequality describing pressures at which the alarm will sound?

Scroll down to find the PSI dot plot for values from 250 to 350.

5. Enter a filter into the *PSI* graph to show the solutions to this inequality. Duplicate the PSI graph.

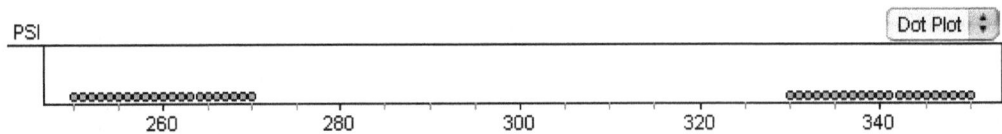

PSI — Dot Plot
260 280 300 320 340

Q9 What filter can you use in the duplicate graph, without an absolute value but with *and* or *or*, that gives the same graph?

Q10 How would you describe how to solve the inequality in the first filter to get the statements in the second filter?

Q11 Compare the Fathom graphs of $\frac{5}{6}\,|\,x - 300\,| \geq 25$ and $\frac{5}{6}\,|\,x - 300\,| < 25$. In your textbook, some graphs are made with filled circles and some are made with empty circles. How does drawing segments with circle endpoints relate to the two Fathom dot plots?

Absolute Value—Radio Contact

Objective: Students will solve inequalities with absolute values and learn about solution sets and the logical meaning of *and* and *or*.

Student Audience: Algebra 1

Activity Time: 25–35 minutes

Setting: Paired/Individual Activity (use **Contact.ftm**)

Mathematics Prerequisites: Students can solve equations.

Fathom Prerequisites: Students can open a Fathom document and read a dot plot.

Fathom Skills: Students will duplicate graphs and edit filters.

Notes: Solutions to these inequalities are represented in Fathom by filters on dot plots representing a number line. As you facilitate students' work, encourage them to develop their own methods for solving inequalities with absolute values. Have students check their results by entering them as filters and comparing the resulting graphs with the solutions for the original inequality. As you circulate among working groups, look for students with unique solutions to Q4, Q7, and Q10. During sharing, ask these students to explain the steps in their methods.

Q1 Most students are familiar enough with time to decide just by thinking (or looking at a clock) that the conversation could have taken place between 2 minutes before the hour and 8 minutes after the hour.

INVESTIGATE

Q2 The numbers are from -2 to $+8$. As needed, help students see that negative numbers indicate times before the hour.

Q3 $(x \geq -2)$ and $(x \leq 8)$; Some students may combine these into the equivalent $-2 \leq x \leq 8$.

4. Students who answered Q3 with $-2 \leq x \leq 8$ will need to change their answer to two inequalities joined by *and*.

Q4 Students will invent their own steps to arrive at the answer. Some may use the distributive property, while others may reverse one of the inequalities when they dissolve the absolute value. Here is one possible sequence.

$$|x-3| \leq 5$$
$$(x-3 \leq 5) \text{ and } (-(x-3) \leq 5)$$
$$(x-3 \leq 5) \text{ and } (x-3 \geq -5)$$
$$(x \leq 8) \text{ and } (x \geq -2)$$

Q5 $|x-3| > 5$ or $\text{abs}(x-3) > 5$; if students use $\geq$ instead of $>$, help them see that endpoints are not included in either set.

Q6 $(x > 8)$ or $(x < -2)$

Q7 Here is one of many legitimate sequences of steps.

$$|x-3| > 5$$
$$(x-3 > 5) \text{ or } (x-3 < -5)$$
$$(x > 8) \text{ or } (x < -2)$$

Q8 $\frac{5}{6}|x-300| \geq 25$

5.

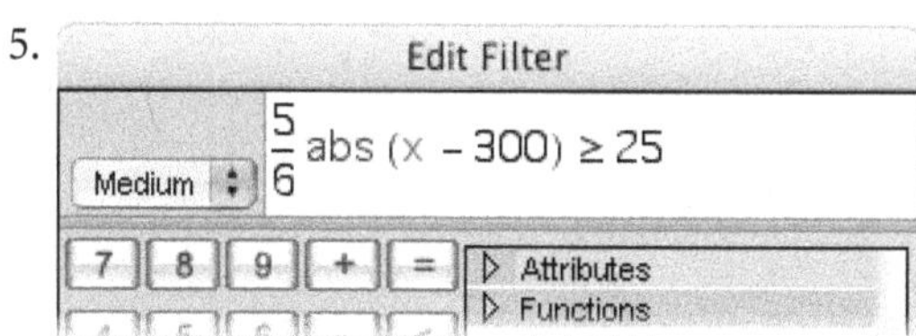

Q9 $(x \geq 330)$ or $(x \leq 270)$

Q10 Here is one sequence of steps.

$$\tfrac{5}{6}|x-300| \geq 25$$
$$|x-300| \geq 30$$
$$x-300 \geq 30 \text{ or } -(x-300) \geq 30$$
$$x-300 \geq 30 \text{ or } x-300 \leq -30$$
$$x \geq 330 \text{ or } x \leq 270$$

The alarm should sound if the pressure is at or above 330 psi or at or below 270 psi.

Q11 The first Fathom graph contains the points 330 and 270, which indicate the closed points. Points that would be open are not included in the second graph.

Two-Variable Inequalities—The Quest

The White Wizard has appointed you to lead a party of elves and trolls on a quest. He has given you the measures of important qualities for potential team members and the total measures required for the quest.

In Endurance, an elf measures 7 and a troll measures 3; you need more than 40.

In Stealth, an elf measures 10 and a troll measures –6; you need more than 9.

In Cunning, an elf measures −3 and a troll measures 5; you need at least 12.

In Diet, or food consumption, an elf requires 0.75 and a troll requires 3; you can provide no more than 25.

Q1 If you gathered a party of four elves and five trolls, what are your total measures for Endurance, Stealth, Cunning, and Diet?

Q2 Create a party that has both Stealth more than 14 and Cunning at least 20. What would Diet be for your party?

INVESTIGATE

1. Open **The Quest.ftm**. You will find a table of every combination of elves and trolls from none to ten of each.

You can add the formula to the table by choosing **Table | Show Formulas.**

2. You want to find the total measure of endurance for every possible party. Add an attribute named *Endurance* and the formula giving 7 for each elf and 3 for each troll: 7 • Number_of_Elves + 3 • Number_of_Trolls.

Instead of typing formulas into the formula editor, you can open Attributes and double-click *Number_of_Elves* and *Number_of_Trolls.*

3. Continue to build your data set by creating attributes and formulas for *Stealth, Cunning,* and *Diet.*

Q3 Find the values of your new attributes for five elves and three trolls.

Q4 Find a party that has both *Stealth* more than 14 and *Cunning* at least 18.

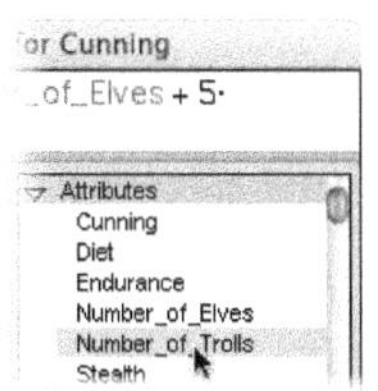

You can find a party that meets all the requirements of the quest by graphing all the possible parties and filtering the parties suitable for the quest from the parties that don't measure up.

4. Create a scatter plot of (*Number_of_Elves, Number_of_Trolls*)

With the graph selected, choose **Object | Add Filter** and enter Endurance > 40.

5. Create the first separation by adding a filter to the graph with the formula Endurance > 40.

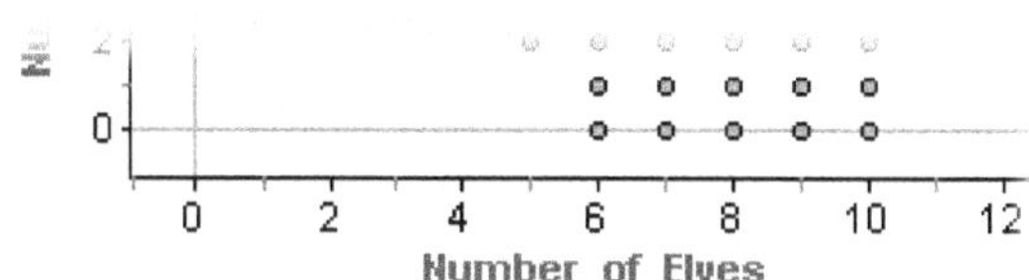

Q5 Describe how the graph looks after you have added the filter. What effect does the inequality sign have on the graph?

To include the equality, hold down Option (Mac) or Ctrl (Win) while clicking > or < to get ≥ or ≤.

6. Duplicate the graph three times and edit filters for *Stealth*, *Cunning*, and *Diet*.

Q6 Select points until you find at least one that is part of the solution set for each of the four graphs. Give a complete description of that point.

Highlighting a point on one graph will highlight that same case in the other graphs and in the table.

Each graph is the graph of an inequality with a boundary and points on one side of the boundary line.

Q7 On paper, start with *Endurance* > 40. Replace Endurance with 7e + 3t and solve for *t*. Replace the inequality with an equal sign. What is the equation?

Phrased another way, the number of elves times 7 plus the number of trolls times 3 gives the total endurance of the party.

7. To see how the line relates to the filtered points, add a plot of the equation to the *Endurance* graph.

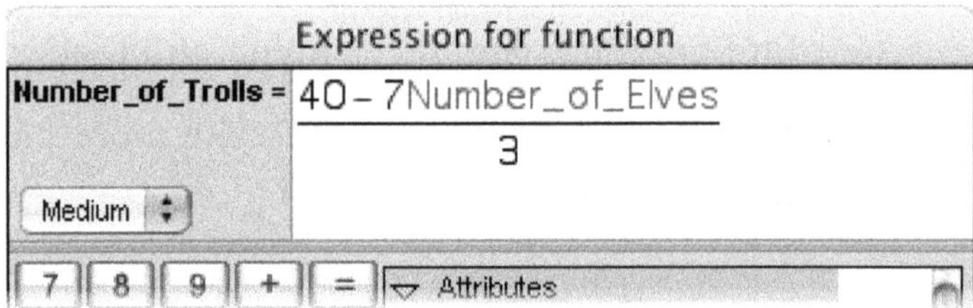

Q8 What is the relationship between this line and the graph?

Q9 Just as you did for *Endurance* in Q7, substitute the formulas for *Stealth*, *Cunning*, and *Diet* and the corresponding inequalities. Replace each inequality with an equal sign and solve for *t*.

8. To see how lines relate to other types of inequalities, add graphs of the boundary lines to each of the other graphs.

Q10 Summarize what you have discovered about graphing inequalities from these graphs. In particular, when are the points on the line part of the solution and when are they not? When do you shade above the line and when do you shade below?

EXPLORE MORE

Create your own requirement and the measures for elves and trolls to further define the possible cases. You may need to experiment with the type of inequality used and the numbers you create so that your graph will contribute to further limiting the possible parties. Explain how your requirement works and the new solution that will satisfy all five of the needs.

Two-Variable Inequalities—The Quest

Objective: Students will translate the inequalities from the problem context and graph them as a filter on a two-dimensional dot plot to see the shading. The inequality is then turned into an equation and the boundary line is graphed.

Student Audience: Algebra 1

Activity Time: 30–40 minutes

Setting: Paired/Individual Activity (use **The Quest.ftm**) or Whole-Class Presentation (use **The Quest Present.ftm**)

Mathematics Prerequisites: Students can solve linear equations in general form for *y*.

Fathom Prerequisites: Students can add attributes and edit formulas, create scatter plots, duplicate graphs, edit filters, and plot functions.

Notes: To help students get the most out of the activity, discuss their methods for answering Q1 and Q2 before they open Fathom. Ask students to share how they found their answers to Q1. If students don't suggest it, show a two-way table similar to this:

	Party		Total
	4 elves and 5 trolls		
Endurance	7×4	3×5	$28 + 15 = 43$
Stealth	10×4	-6×5	$40 - 30 = 10$

As you discuss Q2 you might ask: Is your party the smallest party that meets those qualifications? What other parties meet these qualifications? Ask students to share their answers and describe how they got them. Use the discussion to make sure students understand what they are doing and why. Two-way tables can help students keep track of their work. The attributes have been named, for example, *Number_of_Elves* rather than simply *Elves* so that students will not say "seven elves" and miss the fact that they are to multiply 7 by the number of elves.

As you observe pairs working through the investigation, ask questions that will help students think about what they are doing. For example, once students finish step 6, ask them to describe the relationship between the inequality sign and the location of the points on the graph.

For a Presentation: When you use this as a whole-class presentation, you might want to set the font size to the largest in **Preferences. The Quest Present.ftm** has the formulas for the various measures already entered into the table. Show these formulas as you ask students to discuss what they mean. To reveal the answers to Q3 and Q4, scroll down the table, rather than opening it further; this will save space in the window. Build the first graph (step 4) next to the table. As students discuss the answers to the remaining questions, scroll down for the graphs with the filters, which are also in the document. You might use Q10 as an individual homework assignment.

Q1 For a party of four elves and five trolls, *Endurance* = 43, *Stealth* = 10, *Cunning* = 13, and *Diet* = 18.

Q2 Solutions having ten or fewer elves and ten or fewer trolls are given here as ordered pair (*elves, trolls*), followed by the measure Diet for that ordered pair: (6, 7) 25.5; (7, 9) 32.25; (8, 8) 30; (8, 9) 33; (8, 10) 36; (9, 10) 36.75; (10, 9) 34.5; and (10, 10) 37.5.

INVESTIGATE

Q3 For a party of five elves and three trolls, *Endurance* = 44, *Stealth* = 32, *Cunning* = 0, and *Diet* = 12.75.

Q4 Answers will vary. Solutions are given here as the ordered pair (*Number_of_Elves, Number_of_Trolls*): (5, 6); (6, 7); (6, 8); (7, 8); (7, 9); (8, 8); (8, 9); (8, 10); (9, 9); (9, 10); (10, 9); (10, 10).

Q5 It consists of all the points in the upper right part of the graph. The boundary seems to run from (2, 9) to (6, 0). The inequality is greater than, so the points are to the right of the boundary.

Q6 The party might contain five elves and six trolls for *Endurance* = 53, *Stealth* = 14, *Cunning* = 15, and *Diet* = 21.75. Other parties have four elves and five trolls (*Endurance* = 43, *Stealth* = 10, *Cunning* = 13, and *Diet* = 18) or six elves and six trolls (*Endurance* = 60, *Stealth* = 24, *Cunning* = 12, and *Diet* = 22.5).

Q7 $t = \frac{40 - 7e}{3}$ or $t = \frac{40}{3} - \frac{7}{3}e$

Q8 Most of the points are above or to the right of the line; some of the points are on the line.

Q9 *Stealth:* $t < \frac{10e - 9}{6}$ or $t < \frac{5}{3}e - \frac{3}{2}$

Cunning: $t \geq \frac{12 + 3e}{5}$ or $t \geq \frac{12}{5} + \frac{3}{5}e$

Diet: $t \leq \frac{25 - 0.75e}{3}$ or $t \leq \frac{25}{3} - \frac{1}{4}e$

Q10 When $t \geq$ or $t \leq$, some points will fall on the line, but not when $t <$ or $t >$. When $t >$ or $t \geq$, the points will be above (or on) the line, and when $t \leq$ or $t <$, they will be on or below the line.

Linear Systems—The Road Trip

Ellen and Rachel have been camping for the week in Macinaw City, Michigan. At 8:00 AM, they head for their home in Detroit. At the same time, Ellen's brother, Paul, and his friend Ted leave Detroit, heading north for their week of camping. They plan to meet along the way to shift the camping gear from Ellen's car to Paul's car. As Ellen and Paul drive, Rachel and Ted chat with each other on their cell phones. They exchange information about the mile markers they pass and the time they pass them. At 9:30 AM, they use three-way calling to contact you. They know you'll be sitting at a computer with Fathom, so you can help them predict where they will meet. They relay their data to you.

INVESTIGATE

1. Open **Road Trip.ftm.** You will find a case table of mile markers along I-75 and the times of day recorded as either Paul or Ellen passed them.

Q1 At what time did Paul pass mile marker 100? Estimate what time Ellen passed mile marker 240 (based on observations before and after that marker).

2. The times are in hours and minutes of the time of day. To have a single time value, add a new attribute, *Time,* and enter the formula Time_H + Time_M. Fathom automatically combines and converts these time units to hours. To eliminate the decimals, you might change the units to minutes.

Times of day at mile markers on I75

	Mile_Marker	Time_h	Time_m	Driver	Time
units	miles	hours	minutes		minutes
=					Time_h + Time_m
1	287 mi	8 h	12 min	Ellen	492 min
2	58 mi	8 h	12 min	Paul	492 min
3	63 mi	8 h	15 min	Paul	495 min
4	71 mi	8 h	21 min	Paul	501 min

Q2 What is the value of *Time* when Paul passed mile marker 71? Give a real-world interpretation of this number.

3. Drag a new graph from the shelf into the document and make a scatter plot with *Time* on the vertical axis and *Mile_Marker* on the horizontal axis. Drag the attribute *Driver* to the middle of the graph. Note how the dots change and how a legend appears below the graph.

Q3 What does this graph tell us about Paul's trip?

Q4 Compare the graphs of Ellen's trip and Paul's trip. What do you see?

Each line on the graph has an equation. The equations of the two lines form a system of equations relating the variables *Mile_Marker* and *Time*. A solution to the system consists of the values of distance and time at which the parties will meet—that is, the point at which the lines intersect, which is the very point you are to predict.

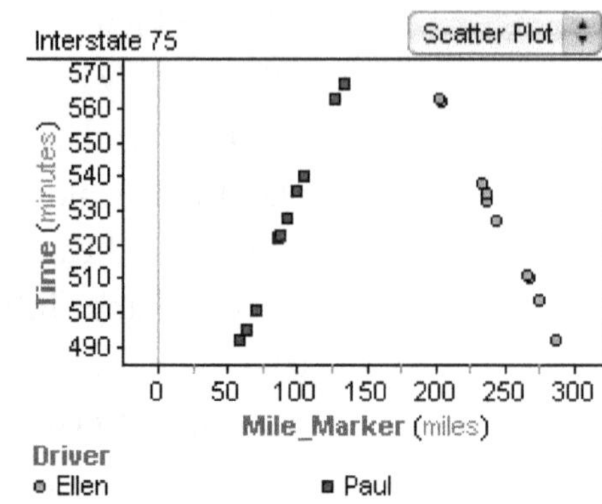

You may have to adjust your graph scale to show the intersection point.

4. Using Fathom objects, estimate the coordinates of the point of intersection.

Q5 Do these coordinates give you enough information to answer the travelers' question? Explain.

5. Drag the variables to the graph again, but this time, plot *Mile_Marker* (vertical) against *Time* (horizontal). Now find the point where the two lines intersect.

Q6 Do these coordinates give you enough information to answer the travelers' question? Explain.

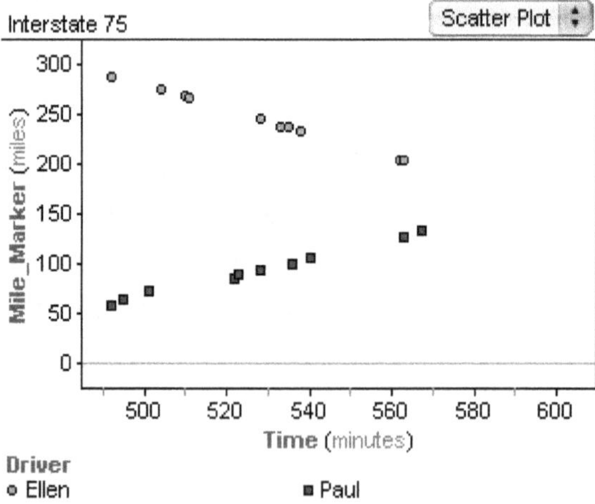

EXPLORE MORE

1. Once you have helped Paul and Ellen identify a meeting place, you wonder about their driving speeds. You notice that one of them is going at a rate very close to 1 mi/min. What is that speed in miles per hour, and which driver is driving at that rate? Is the other driver going faster or slower? What is the average speed of the other driver?

2. For each of the following scenarios, create a graph that fits the given information. Describe the solution(s) to the system for each case.

    a. Paul and Ellen are both driving north on I-75 and both start at the same time. Paul starts about 30 mi south of Ellen but is driving faster.

    b. Paul and Ellen are both driving north on I-75 and both start at the same time. Paul starts about 30 mi south of Ellen, who is driving faster.

c. Paul and Ellen are both driving north and both start at the same time. This time Paul starts about 30 mi south of Ellen and is driving at the same average speed.

d. Paul and Ellen are both driving north and both start at the same time. This time Paul starts at the same point as Ellen and is driving at the same average speed.

3. If possible, graph a system of equations that has no solutions and one that has more than one solution. If either is possible, what equations did you graph? If either is not possible, explain why.

Objective: Students will use the graph of a linear system of equations to model and solve a real-world problem. In Explore More 3, students see systems with no solutions and with infinitely many solutions.

Student Audience: Pre-algebra, Algebra 1

Activity Time: 15–25 minutes

Setting: Paired/Individual Activity or Whole-Class Presentation (use **Road Trip.ftm** for either setting)

Mathematics Prerequisites: Students have general equation-solving skills.

Fathom Prerequisites: Students can add attributes, edit attribute formulas and units, create a scatter plot, use movable lines, and find the coordinates of any point on a graph.

Fathom Skills: Students will add a third attribute to the middle of a graph.

Notes: You might use this activity to introduce systems of equations, because it focuses on understanding what a system of equations is and how a system can be represented and solved graphically. This activity does not include solving a system symbolically. Any approximately correct answer, with proper interpretation, should be accepted. Even though the cars travel at nearly constant speeds, it is not realistic to expect to find the exact time and position they will pass. Students will probably use a pair of movable lines to make their estimates. As you listen in on groups, look for students to share different methods with the class. Teachers have found that students learn a lot from Explore More 2—ask a pair to discuss it during sharing for the benefit of those who may not have completed it. You might suggest to that pair that they ask for class suggestions before graphs are shown.

For a Presentation: As your students discuss their answers to questions such as Q1 and Q2, ask them to explain how they found their answers. For Q4, expect students to have different comparisons to share with the whole class. Explore More 2 and 3 might be assigned to individuals to complete on their own.

INVESTIGATE

Q1 Paul passed mile marker (MM) 100 at 8:56 AM. Ellen passed MM 240 at about 8:50 AM. You might ask further questions such as, How did you find your estimate?

Q2 Paul reached MM 71 at 501 minutes after midnight, or 8:21 in the morning. Some students will divide by 60 and interpret the answer, 8.35, as 35 minutes after 8 rather than $\frac{35}{100}$ of an hour past 8.

Q3 Answers may vary. Paul was traveling at a fairly constant speed (going north).

Q4 Answers may vary. Ellen was also traveling at a constant speed, but she was going south. Ellen is going faster than Paul. It appears they will pass each other around the 150-mile mark at about 600 minutes after midnight.

Q5 The lines intersect at approximately (160, 600). That means that each car will be at MM 160 at 600 minutes after midnight, or at 10:00 AM. You might encourage students to zoom in by rescaling the graph.

Q6 The lines intersect at approximately (600, 160). This gives the same solution as in Q5.

EXPLORE MORE

1. A good line of fit for Paul is

$$-435 \text{ mi} + (1.00 \text{ mi/min}) \cdot \textit{Time}$$

Paul's speed is 60 mi/h. Ellen is going faster, closer to 70 mi/h.

2. a. These two lines will intersect when Paul catches up to Ellen.
   b. These lines will not intersect (for positive values). They will continue to get farther apart.
   c. These two lines are parallel and will never intersect (for any time values). The distance between the two cars will stay the same (30 mi).
   d. These are the same line and will intersect at all points. (In reality, the cars may pass each other many times.)
3. A graph with no solutions will have parallel lines; the equations have the same slope. A graph with more than one solution will have two lines drawn on top of each other; one equation is a multiple of the other.

Systems Solving—High Jump Records

For thirty years, the world records in high jump for both men and women increased consistently, with women's records increasing faster than men's. If the pattern continued, would the records someday be the same?

While systems of two variables can be solved graphically, most systems in the real world contain more than two variables and cannot be solved with a two-dimensional graph. For those problems, there are other solution methods. You will use those methods for two variables and use the graph to check how well they work.

Q1 By substituting, we can use two known facts to conclude a third fact. For example, if Carl owns the book and the person that owns the book wants to sell the book, then what fact can we conclude?

One approach to solving the system of equations is substitution, which is allowed by the transitive property of equality: If $a = b$ and $a = c$, then $b = c$.

Q2 If $y = 3x + 2$ and $y = 5x - 12$, then what does the transitive property of equality allow us to conclude? What is the solution of this equation?

INVESTIGATE

1. Open **High Jump.ftm.** You will see collections and graphs for both men's and women's records for competition high jump since 1970.

Q3 When these data were collected, what was the current record for women, when was it set, and who set it?

To make predictions and comparisons, you need to identify the pattern.

2. Select the graph for the women's records and add a movable line. Position the line to fit the data well.

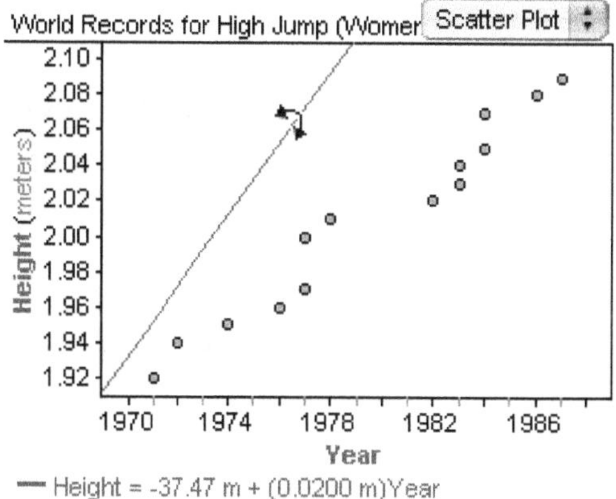

Q4 What is your equation? What does the slope of this graph tell you? At what rate is the women's record increasing?

3. Repeat this process for the men's records.

Q5 What is your line of fit for the men's records? At what rate is the men's record increasing?

Q6 To find where the lines might intersect, use substitution to equate the expressions in the two height equations. Solve this new equation and give a meaning to your solution.

Q7 Substitute your answer to Q6 into each line-of-fit equation you are using from Q4 and Q5. What are the two results, and what do they mean?

4. Add a new graph to the document and choose **Function Plot** from the graph's pop-up menu.

5. Open the graph inspector and set *xLower* to 1970 and *xUpper* to a value more than your answer to Q6. Then set *yLower* to 1.9 and *yUpper* to a value more than your answer to Q7.

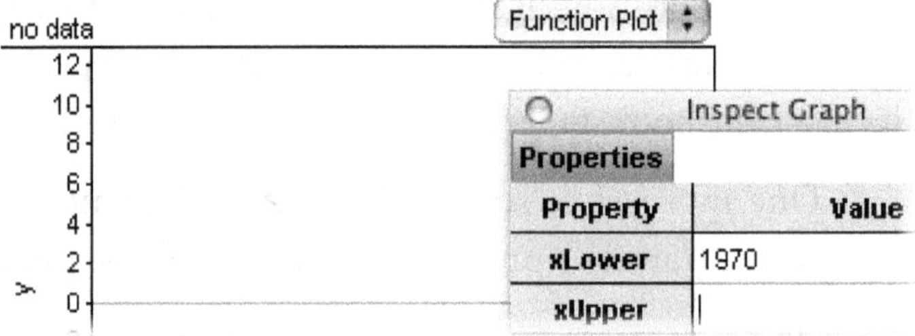

6. Choose **Plot Function** from the **Graph** menu and enter the equation from the women's graph, changing *year* to *x*. Then plot another function using the equation from the men's graph.

Q8 What do you see in this graph? What's the point of intersection?

Q9 Does the point you found make sense? Explain.

EXPLORE MORE

You may have noticed that other students found answers quite a bit different from your answer. These two trend lines have similar slopes. Classmates using only slightly different slopes may get quite different answers. By rotating your movable lines the least bit, you can move the intersection by 50 years or more.

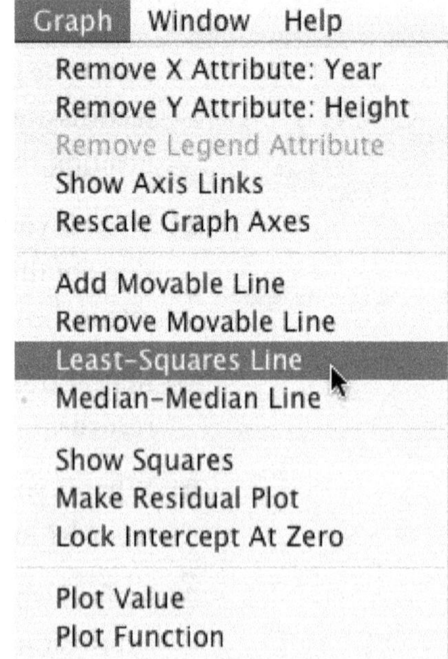

One way to avoid variability between predictions is to have everyone use a specific method to find the line of fit. Fathom has two methods for automatically finding lines of fit. Return to the investigation, but this time do not select the movable line. Instead go to the **Graph** menu and choose either **Least-Squares Line** or **Median-Median Line.** Complete the investigation using these equations.

Systems Solving—High Jump Records

Objective: Students will use substitution to solve a system of linear equations modeling a real-world problem. They will interpret the results and decide on the solution's reasonableness.

Student Audience: Algebra 1

Activity Time: 30–40 minutes

Setting: Paired/Individual Activity (use **High Jump.ftm**)

Mathematics Prerequisites: Students can interpret the slope of a line.

Fathom Prerequisites: Students can add a movable line and create function plots on a graph.

Fathom Skills: In Explore More, students will learn how to add a least-squares line and a median-median line.

Notes: This activity gives students graphic, real-world interpretation of solving a system by substitution. Although the activity explains what substitution is and relates it to other concepts, it does not teach equation-solving steps associated with the substitution method.

As you observe pairs working through the investigation, ask questions that will help students think about what they are doing. You might ask them to explain why the answers for Q7 are the same for both equations and what that means in the context of the problem.

As students share their answers and their reasoning, discuss both the variability in the predictions as well as the unlikelihood that the trend will continue.

Q1 Carl wants to sell the book.

Q2 $3x + 2 = 5x - 12; x = 7$

INVESTIGATE

Q3 Stefka Kostadinova set the record in 1987 at 2.09 m.

Q4 Equations will vary. One good equation is $-18.94 + 0.0106\,Year$. The slope is the average rate of increase (in meters) in the record per year. For women, it is about 0.01 m/yr, or about 1 cm/yr.

Q5 Equations will vary. One good equation is $-14.79 + 0.00866\,Year$. The rate of increase is 0.008 or 0.009 m/yr (8–9 mm/yr).

Q6 Answers will vary depending on equations: For the equations in Q4 and Q5,
$-18.94 + 0.0106\,Year = -14.79 + 0.00866\,Year$
The solution to this equation is year 2139, but other answers may differ by 50 years or more. If the trend over the 30 years in these data continues, this is the year that men and women would have nearly the same record high jump.

Q7 The answers should be the same for both equations. The equations above give 3.74 m, but answers within 25 cm are acceptable. This is the record height both men and women would be jumping in the year 2139.

Q8 The graph shows two nearly parallel lines intersecting at about (2139, 3.74). Student answers will vary by 50 years or more.

Q9 This is a highly unlikely prediction. Just because the trends have been fairly linear for 30 years does not mean they will continue in that pattern for another 150 years (nearly five times as long). In fact, records before 1970 show that the present linear pattern was preceded by a definite nonlinear pattern much longer in length.

EXPLORE MORE

Least-squares: $y = -18.184 + 0.010201x$,
$y = -13.934 + 0.008227x$; (year 2153, record 3.78 m)

Median-median: $y = -21.74 + 0.012x$,
$y = -17.45 + 0.01x$; (year 2145, record 4.00 m)

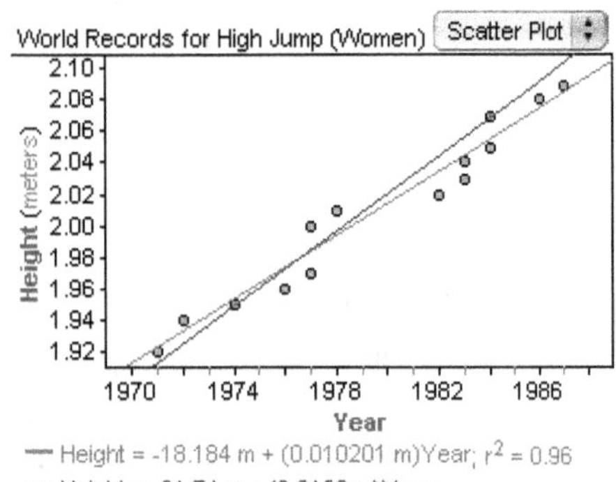

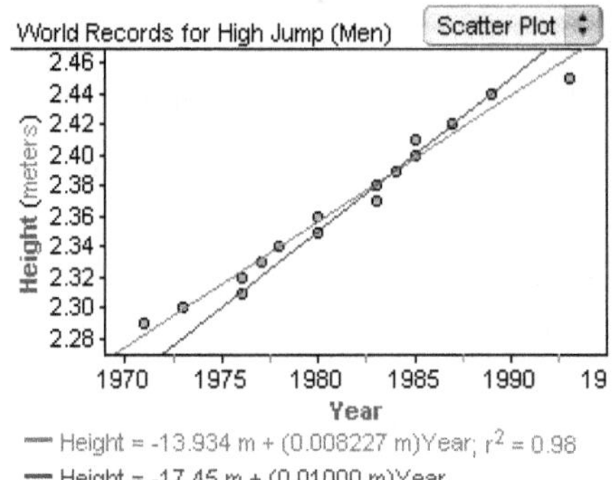

Elimination—Package Charges

You've recently started working as a bicycle delivery person. You've just lost your pricing chart for the third time. Now you are making a delivery and picking up a new package, and you don't know what to charge this customer. You are afraid to call your boss and report that you lost the pricing chart again, as you would probably be fired on the spot.

You know the charge is based on the distance the package must travel and its weight. You remember that the numbers increase by even amounts on the table. The package you just delivered weighed 6 ounces and traveled 4 miles. The charge was \$1.50. You recall your earlier delivery of a 5-ounce envelope that went 8 miles at a cost of \$1.81. This customer is giving you an 11-ounce package that needs to be delivered 3 miles. Do you have enough information to know how much to charge the customer?

To solve this problem, you might set up a system of linear equations. The easiest form for those equations is the general form, $ax + by = c$. If the coefficients of each variable are different in the two equations, an efficient method for solving equations involves adding multiples of equations to find new equations.

Q1 Multiply both sides of the equation $3x + 4y = 5$ by 2. Graph $3x + 4y = 5$ and the new equation. What points on the graph of the original equation does the graph of the new equation pass through? Explain.

Q2 Add both sides of the equations $3x - 3y = -3$ and $2x + 3y = 12$. Graph the two original equations and the equation that is their sum. What points on the graphs of the original equations does the graph of the new equation pass through?

INVESTIGATE

1. Open **Charges.ftm**. You will find a collection of 900 different combinations of weight costs and distance costs (1 to 30 cents per ounce and 1 to 30 cents per mile).

To find the charge for shipping the first package at various weight and distance costs, you multiply 6 ounces times the price per ounce and 4 miles times the price per mile.

Make sure you type the multiplication sign after the 6 and the 4.

2. Add a new attribute with a name like *Package_1* and create a formula for the shipping charges.

Shipping Cost

	Weight_Cost	Distance_Cost	Package_1
=			6Weight_Cost + 4Distance_Cost
1	1	1	10

Q3 By hand, calculate the shipping charge for this package if the cost is 2 cents per ounce and 5 cents per mile. Does your result equal the cost listed in the table?

3. To see what possible weight and distance charges could have been used to determine the charge of 150 cents for the first package, select the scatter plot of all 900 pairs of costs and add a filter Package_1 = 150. The filter removes all the points except those where *Package_1* is 150.

Q4 Choose one point remaining on the graph and explain what it tells you.

Q5 What formula would you use for the cost of the second package, which weighed 5 ounces and went 8 miles?

4. Add another attribute, *Package_2,* and enter the formula for this attribute.

5. To see graphs of both equations, edit the filter to read

(Package_1 = 150) or
(Package_2 = 181)

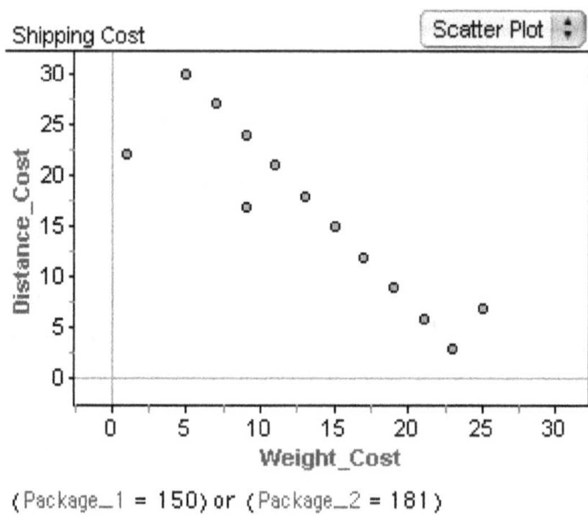

Q6 Use the resulting scatter plot to determine the weight and distance costs. Justify your answer. Then use your solution to calculate the shipping charge for the latest package, 11 ounces across 3 miles.

Until the price list changes, you now know the equation to determine the cost of any shipment. You wonder how you might solve a similar problem without graphing.

To open the slider's inspector, double-click its thumb.

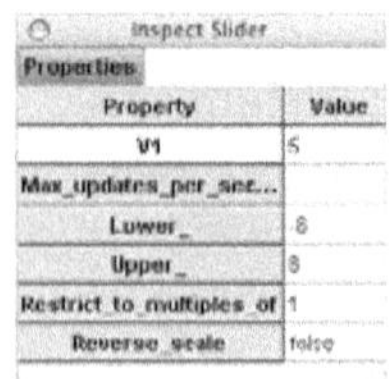

6. Create sliders to use as the equation multipliers. Drag down a new slider, *V1.* Open the slider's inspector so you can set *Lower_* to −8 and *Upper_* to 8. Set Restrict_to_multiples_of to 1. Then select the slider and choose **Object | Duplicate** to create another slider with the same settings.

You'll use graphing to see what multiples to use. You're going to graph a new line by adding the results of multiplying the two equations by the slider values—that is, you'll add these two equations:

$$V1 \cdot (6\,Weight\_Cost + 4\,Distance\_Cost) = V1 \cdot 150$$

$$V2 \cdot (5\,Weight\_Cost + 8\,Distance\_Cost) = V2 \cdot 181$$

More simply, think of adding

$$V1 \cdot Package\_1 = V1 \cdot 150$$

$$V2 \cdot Package\_2 = V2 \cdot 181$$

The sum is

$$V1 \cdot Package\_1 + V2 \cdot Package\_2 = V1 \cdot 150 + V2 \cdot 181$$

In the formula editor, make $\geq$ by holding down Ctrl (Win) or Option (Mac) while clicking $>$.

7. To see both your initial lines and this new line, duplicate the graph by selecting the graph and choosing **Object | Duplicate.** Because some lines may be difficult to see, change the equal sign to greater than or equal to as you edit the filter on this second graph to V1 • Package_1 + V2 • Package_2 ≥ V1 • 150 + V2 • 181. The line itself is the boundary of the dotted region shown on the graph.

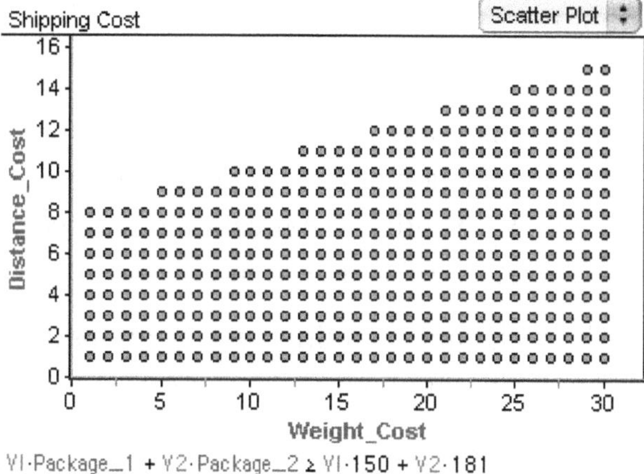

8. To keep track of the actual intersection point, click in the first graph on the solution (17, 12) to turn that point red. Now experiment with the sliders; try positive, negative, and zero values for both. Observe the effects on the second graph.

Q7 What generalization can you make about every line formed by adding multiples of the two equations that you began with?

9. Continue to experiment with the sliders until you have created both a vertical line and a horizontal line. Record the values of *V1* and *V2* that produce each of these lines.

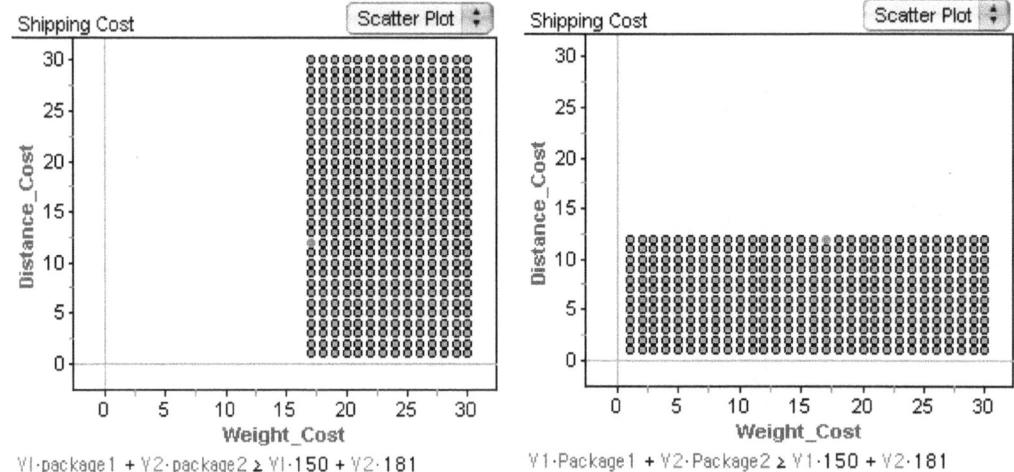

Q8 For the values *V1* and *V2* that you found that create a vertical line, multiply the equation of *Package_1*, which is $6 \cdot Weight\_Cost + 4 \cdot Distance\_Cost = 150$, by *V1*. Multiply the equation of *Package_2* by *V2*. Add the resulting equations together. What do you find? Explain.

This technique of multiplying equations by constants and then adding to eliminate the coefficient of one variable is called *elimination*.

Q9 For the values you find for *V1* and *V2* that create a horizontal line, multiply $6 \cdot Weight\_Cost + 4 \cdot Distance\_Cost = 150$ by *V1* and the second equation by *V2*. Then add the two equations together. Why is this result special?

Q10 Write a summary of the elimination procedure that you can use to solve a pair of linear equations without graphing.

EXPLORE MORE

You may have noticed that other students found different values of *V1* and *V2* to give them vertical or horizontal lines. Continue to look for successful values until you can generalize what is true about any pair that will create a vertical line and any pair that will create a horizontal line. Connect your generalization to the two original equations.

Elimination—Package Charges

Activity Notes

Objective: Students will see that the sum of two linear equations in a system is a linear equation that also includes the solution of the system. (Graphically: The sum of two lines will intersect both lines at the common point.) Using Fathom sliders as equation multipliers, students will discover what the elimination method represents graphically.

Student Audience: Algebra 1

Activity Time: 40–50 minutes

Setting: Paired/Individual Activity (use **Charges.ftm**) or Whole-Class Presentation (use **Charges Present.ftm**)

Mathematics Prerequisites: Students can multiply both sides of an equation by the same number, add two equations, and graph an equation in standard form.

Fathom Prerequisites: Students can add attributes and edit their formulas, add and edit a filter on a graph, use sliders, and duplicate graphs and sliders.

Notes: To decrease the reading, you might want to introduce the scenario and discuss answers to Q1 and Q2 before student pairs begin working. As you listen in on pairs working, make sure they understand what is happening as they create filters for graphs and move the sliders. Student explanations are particularly important for Q8. In Q9, they see that sums that eliminate a variable are graphed as horizontal or vertical lines through that point of intersection; thus, the elimination method can be used to solve a system of linear equations. During discussion, ask students to share their responses to Q7 and Q10.

For a Presentation: Use this activity as a whole-class presentation to introduce elimination as a method of solving simultaneous equations. Or you can use step 6 and following as a short demonstration. For either presentation scenario, ask a student to run the computer as you facilitate the discussion. Before the results of a step are displayed, ask students what they think they will see. **Charges Present.ftm** already includes the formulas for the two known packages. Expand the table to reveal them and make sure students understand why those are the correct formulas. Build the scatter plot for step 3 and discuss Q4 and Q5. The sliders are already built. Scroll down in the document to find them. Allow several students to verbalize their observations before you go on, especially as you discuss Q6 and Q7.

Q1 $6x + 8y = 10$; The graphs are not different. Adding the doubles of two numbers always gives the double of the sum.

Q2 $5x = 9$; The new line intersects the original lines at the point where those two lines intersect.

INVESTIGATE

Q3 $6(2) + 4(5) = 32$ cents; this matches the table value.

Q4 These are all (*Weight_Cost, Distance_Cost*) pairs that would cost $1.50 for the first package. For example, if you charge 7 cents per ounce and 27 cents per mile, then a 6-ounce package going 4 miles would cost $1.50.

Q5 5*Weight_Cost* + 8*Distance_Cost*

Q6 17 cents per ounce and 12 cents per mile; (17, 12) is the point where the two lines appear to cross. Shipping a package weighing 11 ounces to a location 3 miles away would cost $2.23.

7. Note that the change from equality to inequality is not critical, but if you use an inequality you will have to explain that students should concentrate on the boundary line, not all the points above it. If you create the graph with equals, you will find the results a little confusing: The number of points that appear continues to change, because the data set includes only integer points.

Q7 The lines come in all slopes and intercepts, but they always contain the point (17, 12).

Q8 The *Distance_Cost* variable has been eliminated, because the equation of a vertical line does not include the second coordinate. The resulting equation in *Weight_Cost* can be solved to get *Weight_Cost* = 17.

Q9 The *Weight_Cost* variable, the first coordinate in the graph, is eliminated, leaving an equation for a horizontal line, which can be solved to get *Distance_Cost* = 12.

Q10 Multiply equations by constants so that when the results are added, one variable is eliminated. How students decide the constants to multiply by may vary. One way is to multiply the first equation by one coefficient in the second equation and multiply the second equation by the opposite of the corresponding coefficient in the first equation.

EXPLORE MORE

For horizontal lines, the ratio of *V1* to *V2* is always -2, which is the negative reciprocal of the ratio of the corresponding coefficients of *Distance_Cost* in the equations. For vertical lines, the ratio of *V1* to *V2* is always $-\frac{5}{6}$, which is the negative reciprocal of the ratio of the corresponding coefficients of *Weight_Cost* in the equations.

4

Exploring Exponential Equations

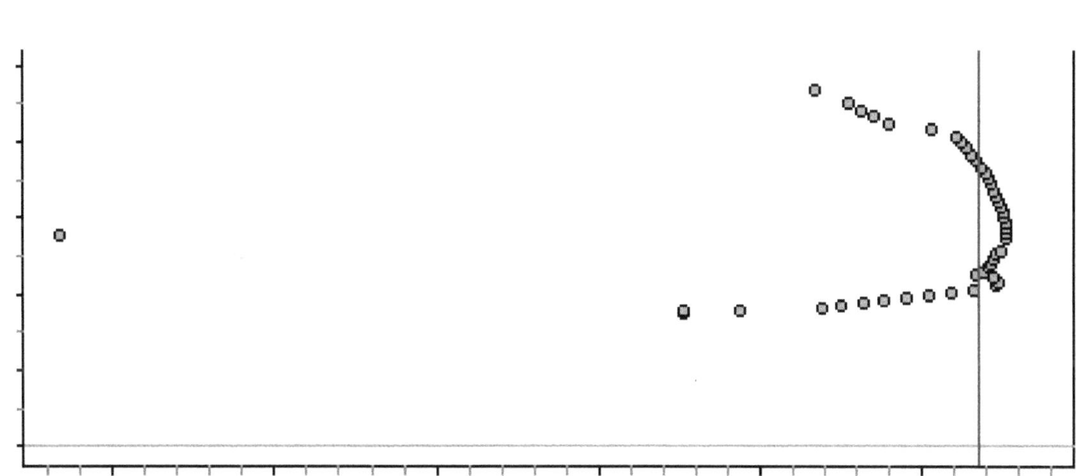

Exponential Growth—Interest

You have $600 that you want to save. You know that it can earn money for you. When you let someone else, such as a bank, use your money, they must pay you rent. That rent is called *interest.*

A dependable and kind relative agrees to pay you interest at a rate of 10%. This means that 10% of the $600 will be added to your savings each year. This is called *simple interest.*

Q1 If you loan your money to your relative, how much will your savings increase each year?

A nearby bank advertises that it will pay you 6% interest on your savings. This seems like a worse deal than your relative's, except the bank's interest is *compounded.* In other words, each year the bank adds to your savings 6% of the growing amount of savings, not just 6% of the original $600.

Q2 If you loan your money to the bank, how much will your savings increase the first year? The second year?

Your goal in this exploration is to decide whether the bank's offer can ever be better than your relative's offer.

INVESTIGATE

To add cases, select the table and choose **Collection | New Cases.**

To show the formula bar, select the table and choose **Table | Show Formulas.**

1. Open a new Fathom document. Set up a case table and make 50 cases. Create attributes for *Year* and for the *Simple_Interest* investment. You can enter the formula for *Year* simply as caseIndex.

2. For *Simple_Interest,* you can use the formula

   prev(Simple_Interest, 600) + 0.1 • 600

   The prev command refers to the previous value of the attribute. The 600 tells Fathom to use the value 600 for the 0th case (which doesn't appear in the table).

Q3 What does adding 0.1 · 600 to the formula accomplish?

3. Create another attribute for *Compound_Interest.* Set up its formula to represent the bank's offer. You can use prev(*Compound_Interest,* 600).

Interest

	Year	Simple_Interest	Compound_Interest
=	caseIndex	prev(Simple_Interest, 600) + 0.1·600	prev(Compound_Interest, 600) + 0.06 prev(Coi
1	1	660	

Q4 What formula did you use to represent the bank's offer?

Q5 Will your savings ever earn more money with the bank's offer than with your relative's?

In non-Fathom situations, you'd like formulas that give you the value of your savings for any year, without having to find the values for all previous years. To help find these formulas, explore with scatter plots.

4. Make a scatter plot of *Simple_Interest* versus *Year*. The data points appear linear, so you'll be looking for an equation of the form $Simple\_Interest = a + b \cdot Year$ 5. Create new sliders, called a and b. Plot the equation $a + b \cdot Year$ on your scatter plot. Adjust a and b to fit the data points as well as possible.

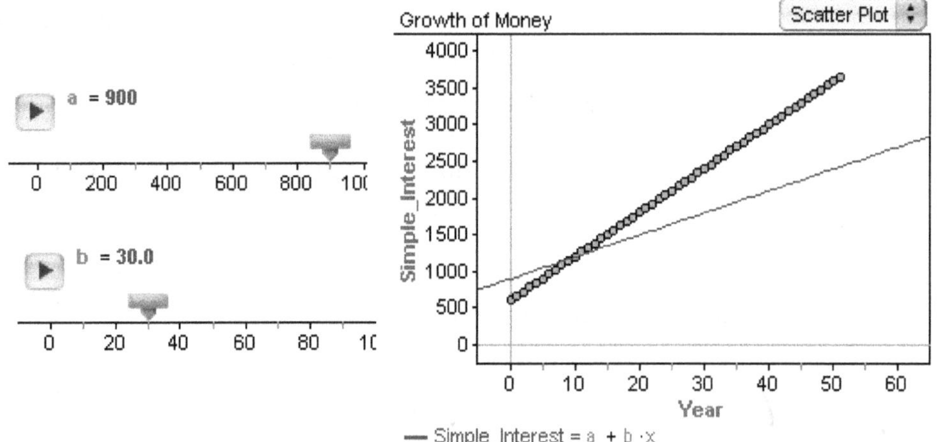

Q6 What values of a and b give the best fit? How do these values relate to \$600 of your initial savings and to the 10% simple interest rate?

5. Drag *Compound_Interest* to be an additional attribute on the vertical axis of your scatter plot.

Q7 Can you adjust a and b to make the graph fit the *Compound_Interest* data? Why or why not?

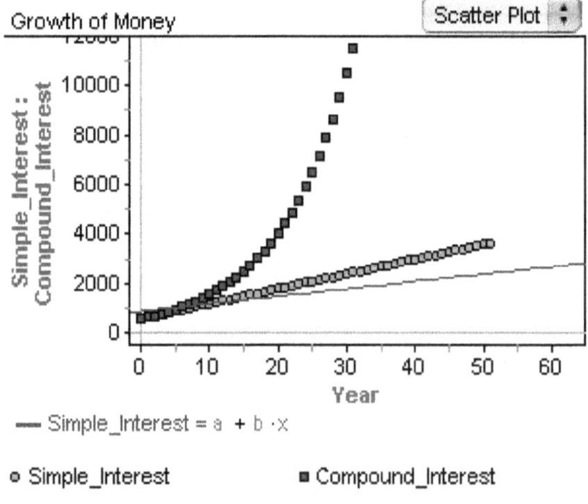

Because compound interest cannot be modeled by a linear equation, you will look for another type of equation that can model it.

6. Perhaps this curve is part of a parabola. Double-click on the plotted equation's formula at the bottom of the *Simple_Interest: Compound_Interest* graph and change it to Compound_Interest = a + b • Year². (To get the exponent, use the ^ key.) Adjust the a and b sliders to match the plotted points as well as possible.

Q8 How well does this graph model the data? Do you think compound interest growth can be modeled by a parabolic graph?

7. Another curve is the graph of an equation in which *Year* is in the exponent. Use Fathom to explore an equation of the form *Compound_Interest* $= a \cdot b^{Year}$. This is called an *exponential equation*. For b, you may need to zoom in on values near 1.

Q9 What values of a and b give the best fit for the data? Do you think compound interest can be modeled by exponential equations?

Q10 When compound interest in this situation is modeled by an equation of the form $A = ab^x$, how do you think a and b relate to the \$600 initial investment and to the rate of 6%?

EXPLORE MORE

1. To check how a and b in the equation *Compound_Interest* $= a \cdot b^{Year}$ are related to the initial investment and to the rate, drag sliders for the initial investment (you might call it P, for *principal*, meaning the initial investment) and for the rate (often called r). Change the formulas for the attributes and for the equation you're plotting to refer to P and r instead of \$600 and the rate. Be sure the graph matches the data points for both compound and simple interest.

2. Often compound interest is calculated and added to the investment more frequently than annually. Suppose \$800 is invested at 12% per year compounded monthly. Each month, $\frac{1}{12}$ of 12% (or 1%) of the investment will be added to the investment. Adjust the collection and graph to model this situation. What is a general formula for the amount under compound interest of an investment of P at an annual interest rate r compounded monthly for t years?

3. What interest rate, compounded annually, gives the same amount of money at the end of a year as the 12% interest rate compounded monthly?

4. Model a situation where \$1 is invested at 171.828% compounded annually. Create a scatter plot showing the growth for eight years. See how well the expression $1 + x$ fits the scatter plot. (You can use x, rather than *Year*, as the variable in the expression, because on a graph, Fathom always understands x to refer to whatever variable is currently on the horizontal axis.) Edit the formula to fit $1 + x + \frac{x^2}{2}$ to the graph. Does this second expression fit more closely? Next use $1 + x + \frac{x^2}{2} + \frac{x^3}{2 \cdot 3}$, $1 + x + \frac{x^2}{2} + \frac{x^3}{2 \cdot 3} + \frac{x^4}{2 \cdot 3 \cdot 4}$, and $1 + x + \frac{x^2}{2} + \frac{x^3}{2 \cdot 3} + \frac{x^4}{2 \cdot 3 \cdot 4} + \frac{x^5}{2 \cdot 3 \cdot 4 \cdot 5}$. Describe how each polynomial expression approximates the amount under compound interest. If the pattern continues, what would the x^6 and x^7 terms be? What do you have to do to get a closer and closer fit?

Objective: Students will learn that a linear equation models simple interest and an exponential equation models compound interest. They will relate the values of a and b in the equations $A = a + bx$ and $A = ab^x$ to the principal and interest rate.

Student Audience: Algebra 1

Activity Time: 35–50 minutes

Setting: Paired/Individual Activity or Whole-Class Presentation

Mathematics Prerequisites: Students can calculate a percentage of an amount, write a percent as a decimal, evaluate equations by substitution, and substitute variable expressions into a formula.

Fathom Prerequisites: Students can set up a case table with cases and formulas, create scatter plots, plot an equation on a scatter plot, and use sliders.

Fathom Skills: Students learn how to write formulas using the prev command.

Notes: As you facilitate student work, probe for understanding of the quantity 1.06, especially in Q10. You might ask pairs who are first to finish the main activity to prepare to present the Explore More questions and answers. Explore More 4 will take the most time for a pair to prepare to present.

For a Presentation: As you lead a class discussion using a presentation computer, emphasize Q6–Q10.

Q1 10% of $600 is $60.

Q2 Year 1: 6% of $600, or $36; Year 2: 6% of $636, or $38.16

INVESTIGATE

Q3 It adds 10% of $600, or $60.

Q4 The formula, partially shown, is

Compound_Interest =
prev(*Compound_Interest*, 600) +
0.06 • prev(*Compound_Interest*, 600)

The algebraic equivalent is

Compound_Interest =
prev(*Compound_Interest*, 600) • (1.06)

Q5 The savings in the bank passes the savings with the relative in year 18.

Q6 $a = 600$, the initial amount; $b = 60$, the amount of interest being added each year. Introduce the terms *principal* for the initial amount of $600 and *interest rate* for the 10%.

Q7 No, the compound interest graph is not linear.

Q8 The compound interest graph is not quite parabolic. It fits well for the first 20 years, but doesn't increase fast enough after that.

Q9 The best value for a is the vertical intercept, 600. The best value for b is 1.06. An exponential equation fits well.

Q10 a is the principal, $600. b is 1 more than the interest rate.

EXPLORE MORE

1. If students use the formula *Compound_Interest* = prev(*Compound_Interest, P*) + r • prev(*Compound_Interest, P*) or its equivalent, then the graph of $P(1 + r)^{Year}$ will pass through the data points wherever the sliders are moved.
2. $A = P(1 + \frac{r}{12})^{12t}$, because there are $12t$ months in t years
3. An interest rate of 12% compounded monthly is roughly equivalent to an interest rate of 12.68% compounded annually.
4. Each successive polynomial expression does a better job of approximating the equation 2.71828^{Year}. The expression $1 + x + \frac{x^2}{2} + \frac{x^3}{6} + \frac{x^4}{24} + \frac{x^5}{120} + \frac{x^6}{720}$ fits better. The expression $1 + x + \frac{x^2}{2} + \frac{x^3}{6} + \frac{x^4}{24} + \frac{x^5}{120}, + \frac{x^6}{720} + \frac{x^7}{5040}$ fits better yet. The number 2.71828 . . . is an important number in mathematics and is called e. It appears in problems having to do with exponential growth.

EXTENSIONS

1. A mortgage is a loan to you for buying a house, so you pay the interest. In some places, such as Canada, a lender telling you the mortgage rate must tell you the annual rate as though the compounding were semiannual (twice a year). If a Canadian lender tells you that the rate is 7.2%, but you want to make payments twice a month, what percent interest will you pay with each payment?

Answer: $\left(1+\frac{0.072}{2}\right)^2=\left(1+\frac{r}{24}\right)^{24}$

$1.036^{\frac{2}{24}}=1+\frac{r}{24}$

$24\left(1.0356^{\frac{1}{12}}-1\right)=r$

$r\approx 0.070838$

The equivalent annual rate is about 7.084% per year or 0.295% twice a month.

You can use Fathom to find the value by creating graphs of two constant functions—$y=\left(1+\frac{0.072}{2}\right)^2$ and $y=\left(1+\frac{r}{24}\right)^{24}$. Use a slider for r and adjust r until the lines coincide.

Exponential Relationships—Population Growth

Governments concern themselves with providing their populations with necessities, such as safe water and food, housing, and medical care. To plan for future needs, they predict their nation's population sizes. Three countries have hired you to help them predict their population sizes in the year 2020.

Q1 What types of things cause a population to change size? Make a list of important factors, beginning with the most influential.

INVESTIGATE

Drag a graph from the shelf, double-click on the collection (but not its name) to open the inspector, and drag attributes from the Eight Countries inspector to the axes of the graph.

1. The first country to hire you is Kenya. Open **Population.ftm** and create a scatter plot of (*Years_After_1948, Population_millions*).

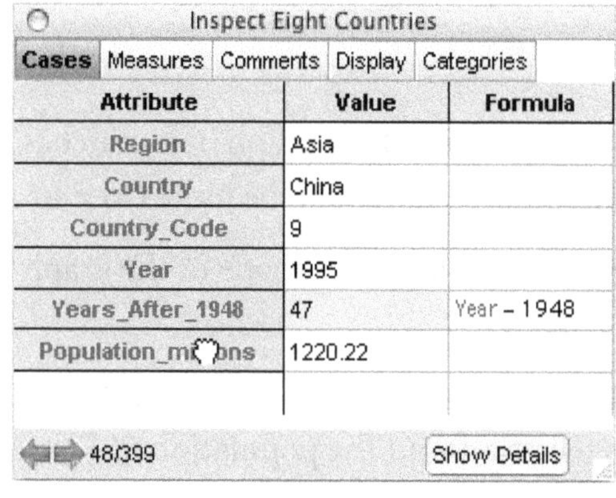

Inspect Eight Countries
Cases Measures Comments Display Categories
Attribute Value Formula
Region Asia
Country China
Country_Code 9
Year 1995
Years_After_1948 47 Year − 1948
Population_millions 1220.22
48/399
Show Details

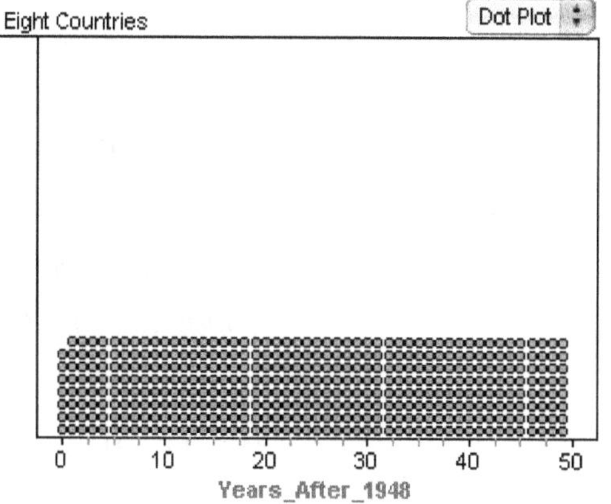

After adding *Population_millions* to the vertical axis, select the graph and go to **Object | Add Filter.** Type Country = "Kenya." Click **OK.** Reselecting **Scatter Plot** as the graph type at the top right will automatically adjust the vertical scale.

2. Add a filter for Country = "Kenya."

The major influence on next year's population is the number of people living in the country today. A good mathematical model for next year's population in many countries involves multiplying the current population by a constant number that represents birth, death, immigration, and emigration. Because of repeated multiplication by this constant number, say b, this model for population growth is exponential and looks like $y = ab^x$. To predict the growth of Kenya's population, you will need to find appropriate values of a, the current population, and b, the rate of growth.

Population is measured in millions as indicated in the attribute name, so we don't include units on the *a* slider.

3. Drag down two sliders and give them names such as a and b (with no units).

Select the graph of Kenya's population and go to **Graph | Plot Function.** Enter a • b ^ Years_After_1948 and click **OK.**

4. Use the values of the sliders to add the graph $ab^{Years\_After\_1948}$ to your scatter plot.

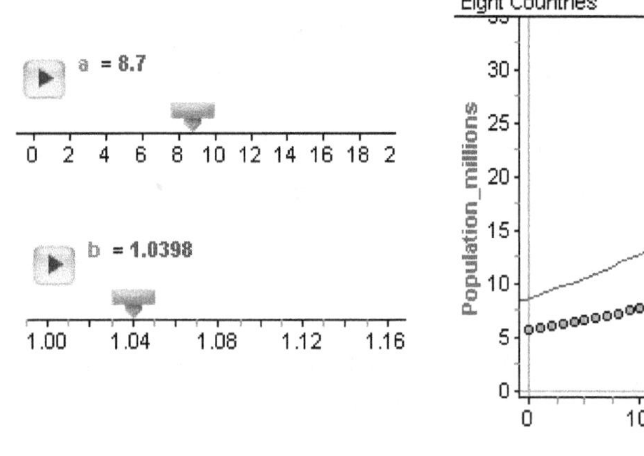

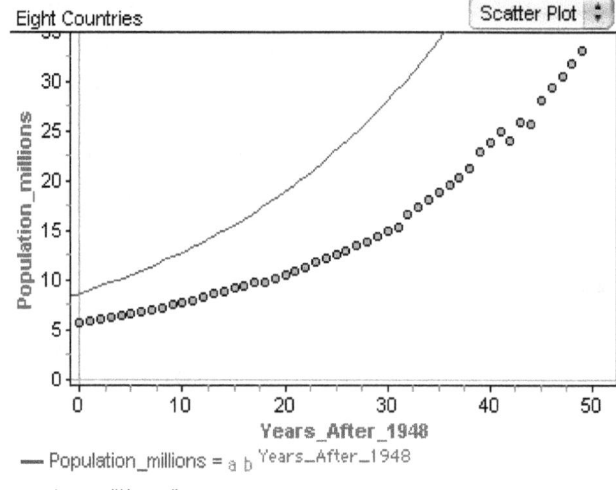

5. Adjust both the *a* and *b* sliders until the graph best matches the data. You will probably want the endpoints of slider *b* to be quite close to 1.

Q2 How does changing the sliders affect the shape of the graph of $Population\_millions = ab^{Years\_After\_1948}$?

Q3 Write the formula for *Population_millions* using your best values for *a* and *b*. What do these values tell you about the population of Kenya?

You may first want to calculate how many years 2020 is after 1948.

Q4 Using these values, what do you predict Kenya's population will be in 2020?

6. The next country to hire you is Israel. Double-click on the filter Country="Kenya" and change "Kenya" to "Israel."

7. Adjust the sliders to match the new scatter plot of Israel's population as well as you can.

Q5 Describe how this graph compares with the graph for Kenya. Use the graph to predict Israel's population in 2020.

8. Because the exponential graph seems to be growing faster than Israel's population, a linear model might be better than an exponential model for this country. Add a movable line to the graph to decide whether this is a better model.

Q6 Where on the graph is the value of *a* represented in each model?

$$Population\_millions = ab^{Years\_After\_1948}$$

$$Population\_millions = a + b \cdot Years\_After\_1948$$

Q7 Give your best model and your prediction for the 2020 population of Israel.

9. The third country to hire you is India. Repeat the process to find a model for India's population.

Q8 What do you predict will be India's population in 2020? Justify your answer.

10. You sometimes need to come up with exponential population models on the fly, without the benefit of Fathom. You can use Fathom to see how to do this. The value of *a* is the initial population. To find the value of *b*, you can use the ratio of the population one year to the population of the previous year. Open the inspector for the collection and enter a new attribute, perhaps called *Ratio,* with the formula next(Population_millions)/Population_millions.

Inspect Eight Countries

Cases | Measures | Comments | Display | Categories

Attribute	Value	Formula
Country_Code	9	
Year	1995	
Years_After_1948	47	Year − 1948
Population_millions	1220.22	
Ratio	1.00972	next(Population_millions) / Population_millions
<new>		

48/399 Show Details

11. To see these values, make a new graph with *Population_millions* versus *Ratio.*

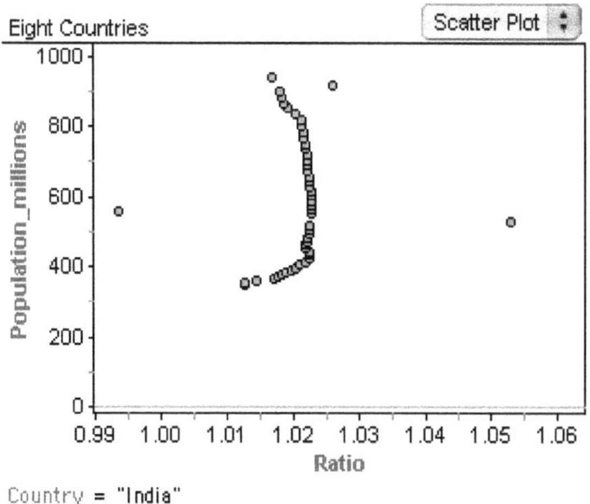

Q9 What can you learn from the fact that these ratios are mostly in a vertical line?

Q10 To see the value more exactly, you might plot the value of median(*Ratio*). How can you use ratios to calculate the value of *b*?

Q11 Carefully explain the role of *a* and *b* in the graph of

$$Population\_millions = ab^{Years\_After\_1948}$$

EXPLORE MORE

1. Use the graphs that you created to determine what happens to the graph of $Population\_millions = ab^{Years\_After\_1948}$ if a or b are negative, if b is 1, or if b is positive but less than 1. Do your observations change how you described the role of a and b in Q11?
2. Suppose the average salary for teachers in one American state is $30,000 and is growing at 4% per year, whereas in a second state, the average salary is more, $40,000, but is growing at a lower annual rate of 2%. Write an equation for each state's average salary. Use Fathom to see when the first state's amount catches up to the second state's.

Objective: Students will use an exponential equation to model population growth and learn how the base is a constant ratio of consecutive values.

Student Audience: Algebra 1, Algebra 2

Activity Time: 45–60 minutes

Setting: Paired/Individual Activity (use **Population.ftm**)

Mathematics Prerequisites: Students can evaluate a formula.

Fathom Prerequisites: Students can use sliders (with units), create a scatter plot and add a movable line, plot a value or an equation, add a new attribute with a formula in the collection inspector, and duplicate a graph and change the variable on an axis.

Fathom Skills: Students learn how to add a filter to a scatter plot and edit a filter.

Notes: As you facilitate student work, watch for students who forget the difference between ab^x and $(ab)^x$ and help them review order of operations for expressions of this type. For groups who finish early, suggest they explore how small changes in a and b affect the prediction.

During sharing after the investigation, students can compare results. Students with almost the same values for a and b can see that their predictions differ substantially and may realize why it's difficult to make accurate predictions.

These data were extracted and the attributes changed slightly from the Fathom sample document found at **File | Open Sample Document | Social Science | World | CountryPopulations.23.ftm.** Together with the document **CountryPopulations.66.ftm,** you should have ample opportunity to extend this activity.

Q1 Answers will vary considerably. The most critical thing is the number of women of child-bearing age. Next is the number of very old and very ill people. This is followed by the prosperity of the country, as well as the ease with which a person can leave or enter the country. All the factors except the last are closely related to the country's current population.

INVESTIGATE

2. The filter provides another way to highlight Kenya's population time series. Because the other countries' data do not appear in the graph, the scatter plot's scale will automatically be determined by only Kenya's data, resulting in a better display when the graph type is reselected. Students could also drag the scales to readjust them. If they are frustrated with this method, remind them of the trick of reselecting the graph type.

5. Finding the best values of a and b is tricky, because it requires fine motor control, the ability to adjust the scale of a slider, and visual estimation skills. If students are ready for a more objective measure of the closeness of the fit, they could go to **Graph | Show Squares** and try to obtain the smallest possible value for the sum of squares. It is possible to get 17.43 (or 17,430,000) for the sum of the squares when $a = 5.285156$ and $b = 1.0374982$. Students will be able to see that the equation matches the data better for smaller values of the sum of the squares, even if they do not understand how it is calculated.

It is interesting to note that one equation fits the points up to about *Years_After_1948* = 32 extremely well, but the curve seems to change at that point. What could have happened in Kenya to make the growth rate increase starting in 1980? How could this graph be defined piecewise?

Q2 Answers will vary. The value of a determines the vertical intercept and the amount of vertical stretch. The bigger the value of b, the faster the graph grows from its starting point of $(0, a)$. Note that both a and b will stretch the curve in different ways. This question should make students more aware of what the values do. To prepare them for later work, you might want to use the formula $y = ab^x$ along with $Population\_millions = ab^{Years\_After\_1948}$.

Q3 Answers will vary but will resemble $Population\_millions = 5.3(1.037)^x$. Kenya's population in 1948 was about 5,300,000. If students have worked with compound interest, you might ask what the base shows about the growth rate. Kenya's population has grown at about 3.7% per year.

Q4 Around 72,500,000 people

Q5 Israel's graph looks more linear. The exponential model predicts about 13,200,000 in 2020, but it may not be very accurate because the exponential equation climbs much faster than the data points do. Although the exponential equation models some countries' populations well, it is not a good model for Israel's population over this time period.

You might mention George Box's maxim: "All models are wrong; some models are useful."

Q6 The values of a are the y-intercepts—in these cases, the populations in 1948.

Q7 The linear model, though not perfect, is probably better than the exponential model and predicts a population of about 8 million people.

Q8 India's formula will be similar to $Population\_millions = 339.2(1.0215)^x$. Because 2020 is 72 years after 1948, the population will be approximately 1.6 billion people. Of the three countries being studied, India seems to have the population best modeled by an exponential equation. The base of the exponential, b, will be close to 1.02, indicating that India's population is growing at about 2% per year.

Q9 The ratios between populations in consecutive years are pretty much constant.

Q10 The median value of the ratios is about 1.02, which is the base of the exponential model.

Q11 The graph of $y = ab^x$ goes through $(0, a)$, and its y-value is multiplied by a factor of b for every increase of 1 unit in x.

EXPLORE MORE

1. If a is negative, the graph of $y = ab^x$ will cross the vertical axis below the horizontal axis and will be a mirror image across the horizontal axis of the corresponding graph with a positive $(y = |a|b^x)$. If $b = 1$, then $b^x = 1$ for all x, and the graph will be $y = a$, a horizontal line through $(0, a)$. If b is positive but less than 1, then the graph goes through $(0, a)$, but the value approaches 0 as x increases.

2. The equations are $Salary = 30{,}000(1.04)^x$ for the first state and $Salary = 40{,}000(1.02)^x$ for the second, where x is the number of years. Students could make a table or use Fathom to graph the two equations. In either case, the salary in the first state exceeds the salary in the second after 15 years.

EXTENSIONS

1. How long did it take the population of Kenya to double from its level in 1948? How long did it take to double again? Explore how long it would take for populations growing exponentially to double at annual growth rates of 2%, 3%, 4%, 6%, 9%, 12%, and 24%. If the annual rate is twice as big, does it take half as long? Look for patterns.

   Answer: The population of Kenya doubles about every 19 years.

Population growing at	Doubles in
2%	34 years
3%	23 years
4%	17 years
6%	12 years
9%	8 years
12%	6 years
24%	3 years

2. Suppose an insurance company estimates that the value of a car decreases by 30% per year. What is the value of a $25,000 car after 4 years? After 10 years? What is the formula for the value after x years? What are some reasons why this model may not reflect reality?

   Answer: About $6000 after 4 years; about $650 after 10 years; $value = 25{,}000(0.7)^x$.

Inverse Variation—Boyle's Law

The amount of time a scuba diver can safely spend under water depends on the amount of air in the diving tank. Your friend has asked for your help in determining how much air is in a particular tank.

You can see that the volume of the tank is 0.340 cubic feet. But you also know that the tank is pressurized to pack more air into a smaller space. As the air leaves the tank, it expands. You can see that the tank is full and that the pressure gauge reads 3500 pounds per square inch (psi). To determine how much air the tank will supply while diving, you need to know what the volume of the air will become as it is released from the tank and the pressure becomes the normal air pressure of 14.7 psi.

INVESTIGATE

1. You have found some data on the Internet and put it into a table to help you look for patterns. Open the document **Boyle.ftm** and create a scatter plot of *Volume* versus *Pressure.* The data are not linear, but perhaps they're exponential.

By highlighting the point representing volume 48, you can read the corresponding pressure from the status bar. You can also look in the table.

Q1 As the volume is halved from 48 to 24, what happens to the pressure?

Q2 Does the pressure approximately double when the volume goes from 30 to 15? Does the pressure always double when the volume is cut in half? Show the evidence for your conjecture.

Q3 If the equation were exponential, would the pressure double when the volume is cut in half?

2. Because the data don't appear to be exponential, you decide to try modeling with a power equation of the form $y = k \cdot x^b$. Unlike in an exponential equation, the *Pressure* attribute will be in the base rather than in the exponent. Add sliders for k and b and add the graph of the power equation to your scatter plot.

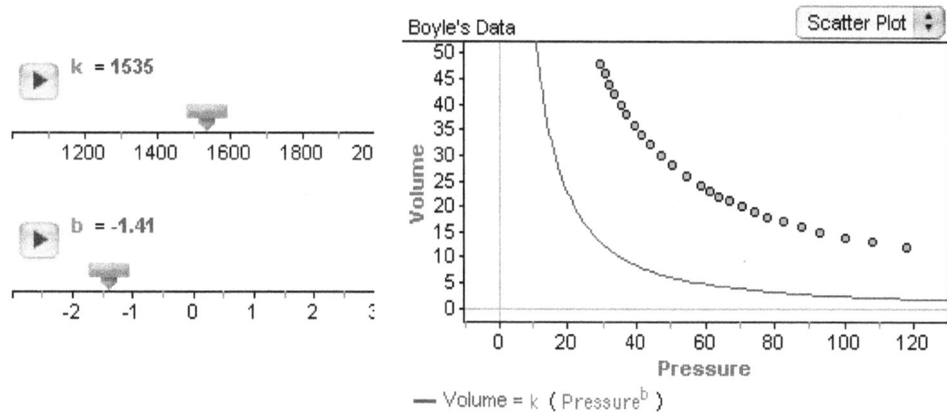

3. Adjust the sliders to fit the data as well as you can.

Q4 According to your model, what pressure is needed to reduce the volume below 10 ft^3?

You may have found that a good value for b is -1. It can be simpler to write the equation as $Volume = k\left(\frac{1}{Pressure}\right)$. The coefficient k of this equation is called the *constant of variation.* Note that $k = Pressure \cdot Volume$. Finding the value of this constant in particular situations is important for answering pressure-volume questions like Boyle's.

4. You can see this relationship in the data table by adding a new attribute, say, *Constant,* with the formula *Pressure · Volume.*

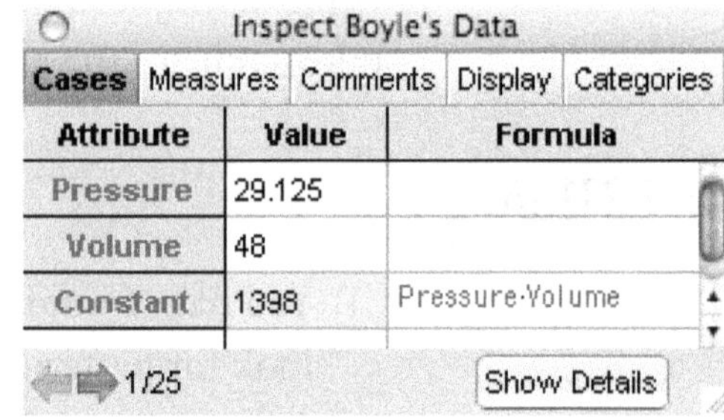

To see a pattern in these values, look at a dot plot of *Constant* and plot the value mean(Constant).

Q5 How does this mean relate to the values of the sliders? Explain.

Q6 Use what you've learned about the data table and return to Boyle's situation. What's the constant of variation for the scuba tank you're examining? Why might it be different from the constant of variation in the table?

Q7 What volume will the air have when it is released from the tank and the pressure becomes 14.7 psi?

EXPLORE MORE

1. To explore more equations of the form $xy = k$, or the equivalent $y = k\left(\frac{1}{x}\right)$, create a new function plot and plot the equation $y = k\left(\frac{1}{x}\right)$. Explore the graph for values of slider k near 0. Describe the graph when $k > 0$, $k = 0$, and $k < 0$.
2. In Explore More 1, you studied a graph in which the *product* of x and y was the constant k. Now explore graphs in which the *sum* of x and y is the constant k. Be as specific as you can about what is always true about the graph and what changes as k changes.
3. What if the *difference* is always k? What if the *quotient* is k? Are you surprised by the results?

Objective: Students will model inversely proportional quantities with equations of the form $y = k\left(\frac{1}{x}\right)$, $xy = k$, and $y = kx^{-1}$. They will investigate the change in one variable as the other doubles, relate $y = k\left(\frac{1}{x}\right)$ to a linear model, and explore the graph of $xy = k$.

Student Audience: Algebra 1, Algebra 2

Activity Time: 30–40 minutes

Setting: Paired/Individual Activity (use **Boyle.ftm**)

Mathematics Prerequisites: Students can rearrange and evaluate a formula.

Fathom Prerequisites: Students can plot a function on a scatter plot, create sliders, change the scale of sliders, and edit slider formulas.

Background: Boyle's law states that if a gas is kept at constant temperature, the pressure and volume are inversely proportional, or have a constant product. Robert Boyle published his findings that pressure times volume is constant in his 1662 article "A Defense of the Doctrine Touching the Spring and Weight of the Air."

Notes: As you facilitate student work, look for students who have complete answers to Q2 and Q3 and have them share their results with the class. If some students are ready to use **Graph | Show Squares** in step 3, ask them to share with the class how they used this Fathom feature to adjust their slider. A variety of answers to Q6 can also be shared; students taking physics will know that temperature is an important factor.

INVESTIGATE

1. These data are Robert Boyle's original. The Internet source is given in the **Comments** panel in the collection inspector. Robert Boyle's methodology and his published findings are intriguing reading.

Q1 The pressure roughly doubles from 29.125 psi to 58.8125 psi.

Q2 As the volume goes from 30 ft^3 to 15 ft^3, the pressure roughly doubles from 47 psi to 93 psi. Some students might say pressure is not quite doubled as volume goes from 15 ft^3 to 30 ft^3. The following table demonstrates that the ratio of the pressures is very close to 2 in every case:

Volume (ft^3)	Pressure (psi)	Volume (ft^3)	Pressure (psi)	Ratio
48	29.125	24	58.8125	2.019313
46	30.5625	23	61.3125	2.006135
44	31.9375	22	64.0625	2.005871
42	33.5	21	67.0625	2.001866
40	35.3125	20	70.6875	2.00177
38	37	19	74.125	2.003378
36	39.3125	18	77.875	1.980922
34	41.625	17	82.75	1.987988
32	44.1875	16	87.875	1.988685
30	47.0625	15	93.0625	1.977424
28	50.3125	14	100.4375	1.996273
26	54.3125	13	107.8125	1.98504
24	58.8125	12	117.5625	1.998937

Q3 No. If the equation were exponential, the pressure would double for constant changes in the volume, not for proportional changes.

3. Finding the best values is tricky. It requires fine motor control, the ability to adjust the scale of a slider, and visual estimation skills. If students are ready for a more objective measure of the closeness of the fit of their equation to the data, they could go to **Graph | Show Squares** and try to obtain the smallest possible value for the sum of squares. A small sum of squares is 0.2861 when $k = 1407.94$ and $b = -1$. Students will be able to see that the equation matches the data better for smaller values of the sum of the squares, even if they do not understand how it is calculated. This model matches the graph remarkably well, which might make you wonder whether Boyle's data were "fudged." There is some evidence that other historically important data, such as Fisher's genetic data, can be statistically shown to be too close to the predictions to have been produced by experimentation.

Q4 A pressure of 141 psi will decrease the volume below 10 ft^3.

Q5 If the value of k is set to the mean value, 1408.09, the graph goes through the data points.

Q6 $(0.34)(3500) = 1190$. Possible reasons the constant may differ from that of the table include differences in temperature and the nature of the gas being compressed.

Q7 About 81 ft^3

EXPLORE MORE

1. When $k = 0$, the graph is the x-axis, with the point $(0, 0)$ removed (because $\frac{1}{0}$ is undefined). When $k > 0$, there will be two branches of the graph, one in the first quadrant and one in the third. When $k < 0$, the branches are in the second and fourth quadrants. This graph is called a *hyperbola.* Its branches approach but never cross its *asymptotes,* which, in this case, are the x- and y-axes.
2. The equation is $y = k - x$, so the graph is a straight line with y-intercept k and slope -1. It crosses the y-axis above the origin when $k > 0$, at the origin when $k = 0$, and below the origin when $k < 0$.
3. If the constant difference equals $x - y$, then $y = x - k$; if the constant difference equals $y - x$, then $y = x + k$. The graph of each is a straight line with slope 1; k is either the y-intercept or its opposite. If the constant quotient equals $\frac{x}{y}$, then $y = \frac{1}{k}x$; if the constant quotient equals $\frac{y}{x}$, then $y = kx$. In both cases, the graph is a line through the origin with slope either k or its reciprocal.

EXTENSIONS

1. Research the pressure experienced by divers. How deep would a diver need to be in order to be subject to twice the pressure experienced at sea level?

   Answer: A diver at depth 10.3 m under water experiences a pressure of about 2 atmospheres (1 atm for the air and 1 atm for the water). Wikipedia (en.wikipedia.org/wiki/Atmospheric_pressure January 2006) is a source students can visit to learn more.
2. A data analyst uses Fathom to fit the exponential equation $Volume = 46.41(0.98032)^{Pressure-29}$ to the data in the table. Make a convincing argument for why either the exponential or the power model is better. You may want to discuss the vertical intercept and its role in the model, as well as the halving time for volume.

   Answer: $Volume = 46.41(0.98032)^{Pressure-29}$ goes through the point $(29, 46.41)$, which is approximately *Pressure* and *Volume* in the first case. The volume is decreasing approximately 2% per increase of 1 unit of pressure at the beginning of the experiment, but it decreases more slowly in the later values. Not only does a graph of the form $PV = k$ fit the points better, but it also ensures that the graph has no vertical intercept (as pressure decreases to 0, volume expands to infinity). The exponential model implies that when there is no pressure, the volume will be about 82.6, which is physically incorrect. In the activity, we noticed that the volume is halved as the pressure is doubled in Boyle's model, as opposed to having a fixed halving period, as with the exponential function model.

Exponents—Moore's Law

In 1965, a founder of Intel Corporation, Gordon Moore, noted that the number of transistors on a computer chip seemed to be doubling every year and a half. His observation became known as Moore's law. It's been more than 40 years since his statement. Has that rate of growth in transistors been maintained?

INVESTIGATE

1. Open the document **Moore.ftm.** Create a scatter plot of (*Years_Since_1971*, *Transistors*).

Hint: What is the value of a when $x = 0$?

Q1 If you are modeling these data with an equation of the form $y = ab$, what would be a value for a?

2. Add sliders for a and b and plot an equation of the form $y = ab^x$ on the scatter plot. Set your value for a from Q1 and adjust the b slider until the curve fits the points as well as possible.

Q2 According to your model, what is the annual rate of growth of the number of transistors?

Q3 Look at the case table or work with your equation to discuss whether Moore's law is a reasonably accurate description of growth in the number of transistors or their silicon equivalent over time.

The data table doesn't show the years before 1971, so we don't know the value of a for Moore's first observation. Let's say that a equaled 1 c years before 1971. Your base equation can be thought of as $y = b^{c+x}$ as well as $y = ab^x$.

3. Drag a new slider from the shelf and name it c. Keeping the same value for slider b, add the plot of the equation b^(c + x) to your graph. Adjust c to fit the data (and the first equation).

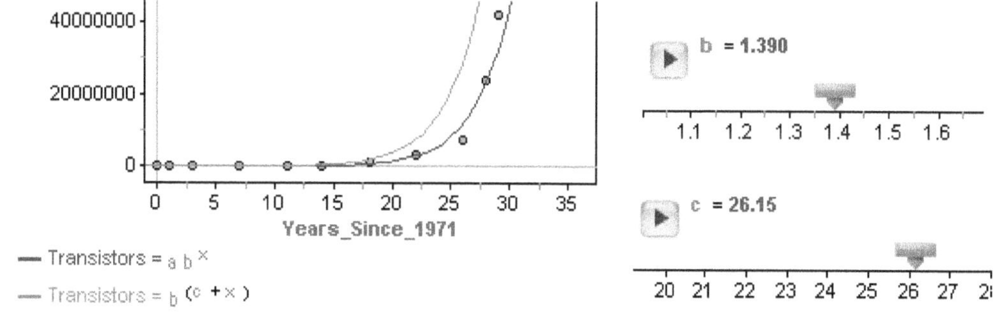

Hint: How many years between 1971 and 2010?

Q4 Which value of c makes the two equations coincide? Using this new model, what is your prediction about the number of transistors on a chip in 2010?

Q5 Your graph seems to show that ab^x and b^{c+x} are equivalent for some values of a and c. Express a in terms of b and c.

Hint: The calculation will involve only one operation and that operation is the main operation in the expressions.

4. To check your conjecture about how a is calculated, make a formula for the a slider in terms of b and c.

Q6 When you adjust the b and c sliders, a new value for a will automatically be generated. Do the two equations continue to coincide as you adjust b and c? What does this tell you?

Q7 Substitute your expression for a into the equation $ab^x = b^{c+x}$. The resulting equation is a rule for working with exponents. State this rule in words.

Other data can be modeled with these equations.

5. Scroll down to the radiosonde data collected as a weather balloon was released into the atmosphere. As the balloon rose, the atmospheric pressure fell.

Hint: Read the collection inspector's **Comments** panel.

Q8 What was the starting height of the balloon? Why did the height not start at 0 km?

As with the Moore's law transistor data, the equation $y = ab^x$ can be used to model the weather balloon situation.

6. Create and adjust the a and b sliders to fit the graph of *Pressure* versus *Height* as well as possible.

Q9 Write the equation of the curve using the values for a and b obtained. What do you think these best values for a and b mean in this situation? What is the percentage drop in pressure per kilometer?

Hint: Try starting the c slider with the values shown.

7. Repeat the procedure outlined in steps 3 and 4 to verify that an equation of the form $y = b^{c+x}$ can also be used to model this situation.

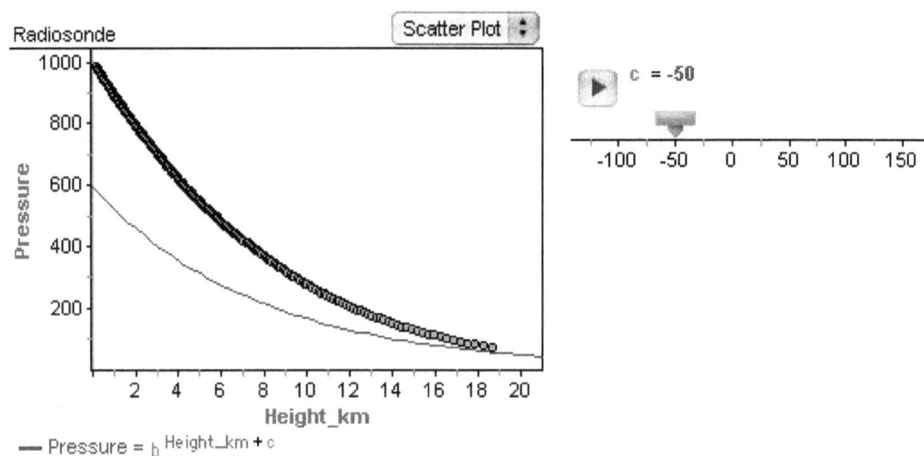

Q10 Which value of c best fits the graph? In this decreasing situation, can a be calculated as it was in step 5? What can you conclude about exponents?

EXPLORE MORE

1. Model the data about Moore's law and the weather balloon data using an equation of the form $y = ab^{-x}$ and compare your results with the model you created in the activity, which used $y = ab^{x}$. How do the values of a compare? How do the values of b compare? Use your observations to write $y = 1000(0.5)^{x}$ in the form $y = ab^{-x}$.

2. You will recall that the $y = b^{c+x}$ model was used with data far from the axis. Use the model written $Transistors = b^{Year-h}$, with the same value of b, to fit (*Year, Transistor*) instead of using *Years_Since_1971*. Can you give some real-world context to the value of h?

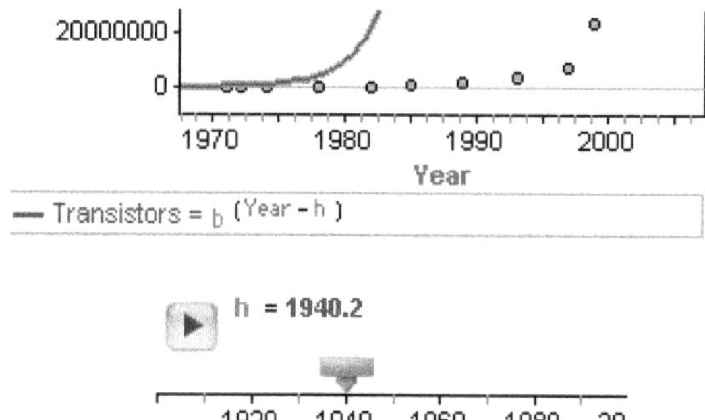

Objective: Students will see the multiplication property of exponents, $b^c \cdot b^x = b^{c+x}$, in a real-world situation. Using Fathom graphs and slider-based equations, students will model real-world data and see that the same data can be modeled two different ways. The multiplication property is the result of connecting the two models. Students will deepen their ability to interpret the role of b in $y = ab^x$ in terms of a percentage increase or decrease.

Student Audience: Algebra 1, Algebra 2

Activity Time: 50–75 minutes

Setting: Paired/Individual Activity or Whole-Class Presentation (use **Moore.ftm** for either setting)

Mathematics Prerequisites: Students can substitute an expression for a variable in a formula and solve an exponential equation by trial.

Fathom Prerequisites: Students can interpret a collection's inspector, use sliders, plot a function on a scatter plot, and determine the value of a slider based on a formula.

Notes: This activity can be broken into two parts by exploring the Moore's law data first and the weather balloon data on another day, or you might use the Moore data in a whole-class presentation and let students work through the weather balloon data in small groups. As you visit groups or during sharing with the whole class, encourage discussion and ask students what they base their opinions on. Questions such as Q7 and Q9 help students understand and communicate mathematics.

For a Presentation: Before you complete a graph or move a slider to a new location, ask students what they expect to happen. Continue to ask questions as you build the document from the data. Many of the questions can start good discussions. Don't give the hint to Q1 unless students need it. (Wait until the count of 40 before you jump in to provide a hint or answer.)

INVESTIGATE

Q1 $a = 2250$, the value when $x = 0$

Q2 Because $Transistors = 2250(1.39)^{Years\_Since\_1971}$, the number of transistors will be 1.39 times as big every year, which is equivalent to a growth rate of 39%.

Q3 The table appears on the next page. Students might look in the case table for years that differ by some multiple of 3 and notice that in all but one case the growth is slower than predicted.

Or they could compare the table with predictions given by the formula in Q2 and see that in 14 years (1985), the number of transistors is only $\frac{1}{5}$ the predicted 1.5 million, and in 29 years (2000), it is only $\frac{1}{50}$ the predicted 1.5 billion.

Alternatively, they might solve $2 = (1.39)^t$ by trial and see that $t \approx 2.1$, which implies that the doubling period is closer to 2 years than to 18 months.

Some students will see the prediction as not very far off considering how long the exponential trend has continued.

3. Students might benefit from setting the slider's endpoints as in the picture.

Q4 Students should get a value of c between 23 and 24, and the two graphs should coincide exactly. The prediction for 2010 would be a chip with between 800 million and 1 billion transistors. If students give answers accurate to the nearest chip, you might mention that the numbers used in the problem are only accurate to three digits, so estimates should be stated only to the nearest million. Students who use 1.38 instead of 1.39 will have an estimate of about 600,000,000.

Q5 Students should find that $a \approx b^c$; for example, $1.39^{23.5} \approx 2295$, a value near 2250. You may need to suggest that students think about which operation besides multiplication and addition is involved in ab^x and b^{c+x}.

4. Once b^c is entered as the formula for a, the a slider can no longer be dragged and its thumb changes shape. The calculated value for a should be reasonably close to 2250.

Q6 The two graphs should coincide exactly, meaning that when $a = b^c$, $ab^x = b^{c+x}$ for every value of x.

Q7 $b^c b^x = b^{c+x}$. The rule is that when powers with the same base are multiplied, the exponents are added. Be alert to wording such as "multiplying the base times itself." b^c means b is used as a factor c times, not that *b* is multiplied by itself *c* times. To see the difference, think of b^1—*b* is not multiplied by itself once.

Q8 The balloon started at a height of 0.227 km above sea level, because it was launched from land (Kansas) that is not at sea level.

The time is interesting. It looks as though a measurement was made 33 s prior to launch and 10.2 s after the launch.

Q9 Student answers will vary but should be similar to $Pressure = 1070(0.88)^{Height}$. The value of *a* in this case indicates that the pressure at sea level is approximately 1070 millibars, and the value of *b* indicates that the pressure is dropping approximately 12% per kilometer (because $0.88 = 1 - 0.12$).

Q10 This situation works exactly like the situation in Q7, with *c* taking on a value close to -54.6. Therefore, $b^c b^x = b^{c+x}$ when *c* is positive or negative and when *b* is greater than or less than 1 (but still positive).

EXPLORE MORE

1. Both $Transistors = 2250(1.39)^{Years\_Since\_1971}$ and $Transistors = 2250(0.72)^{-Years\_Since\_1971}$ model the Moore's law data. Also, $Pressure = 1070(0.88)^{Height}$ and $Pressure = 1070(1.14)^{-Height}$ model the balloon data. To change from one form to the other, keep the *a*'s the same and take the reciprocal of the *b*'s. Here $1.39 \cdot 0.72 \approx 1$ and $0.88 \cdot 1.14 \approx 1$. Numbers that multiply to 1 are called *reciprocals.* It may surprise some students that the amount that one base is greater than 1 is not the same as the amount that the corresponding base is less than 1. We see a similar phenomenon in exchange rates. If an American dollar is worth 1.30 Canadian dollars, the Canadian dollar is not worth U.S.$0.70 but rather U.S.$0.769 because the reciprocal of 1.3 is 0.769. Another form of $y = 1000(0.5)^x$ is $y = 1000 \cdot 2^{-x}$, because 0.5 and 2 are reciprocals.

INVESTIGATE Q3

Year	Transistors	Comparison Year	Difference (in years)	Predicted Number of Doublings	Predicted Growth Factor	Number of Transistors in Comparison Year	Actual Growth Factor	Actual Compared to Predicted
1971	2,250	1974	3	2	4	5,000	2.2	less
1982	120,000	1985	3	2	4	275,000	2.3	less
1997	7,500,000	2000	3	2	4	42,000,000	5.6	more
1972	2,500	1978	6	4	16	29,000	11.6	less
1993	3,100,000	1999	6	4	16	24,000,000	7.7	less
1985	275,000	1997	12	8	256	7,500,000	27.3	less
1974	5,000	1989	15	10	1,024	1,180,000	236.0	less
1978	29,000	1993	15	10	1,024	3,100,000	106.9	less
1982	120,000	1997	15	10	1,024	7,500,000	62.5	less
1985	275,000	2000	15	10	1,024	42,000,000	152.7	less
1989	1,180,000	2004	15	10	1,024	125,000,000	105.9	less
1971	2,250	1989	18	12	4,096	1,180,000	524.4	less
1982	120,000	2000	18	12	4,096	42,000,000	350.0	less
1972	2,500	1993	21	14	16,384	3,100,000	1,240.0	less
1978	29,000	1999	21	14	16,384	24,000,000	827.6	less
1972	2,500	1999	27	18	262,144	24,000,000	9,600.0	less
1974	5,000	2004	30	20	1,048,576	125,000,000	25,000.0	less
1971	2,250	2004	33	22	4,194,304	125,000,000	55,555.6	less

2. The value of h is between 1947 and 1948. In 1947, the first transistor was produced, and UNIVAC, the first commercial computer, followed in 1948.

EXTENSIONS

1. If $ab^{-x} = b^{-(x+c)}$ for all values of x, what is a in terms of b and c?

   Answer: Applying exponent laws yields

   $ab^{-x} = b^{-x-c}$

   $ab^{-x} = b^{-x}b^{-c}$

   Because the factor of b^{-x} is common to both sides,

   $a = b^{-c}$ or $\frac{1}{b^c}$

2. Dynamically fit an exponential equation with base 2 to the Moore's law data to determine the length of time required to have the number of transistors double. Compare this length of time with the 18 months prediction.

   Answer: Taking the earlier expression, $Transistors = 2250(1.39)^{Years\_Since\_1971}$; rewriting it in base 2 will yield something similar to $Transistors = 2250(2)^{0.475 \cdot Years\_Since\_1971}$.

   For the number of transistors to double, the exponent 2 must go up by 1.

   $Years\_Since\_1971 \approx \frac{1}{0.475}$

   $Years\_Since\_1971 \approx 2.1$

   It seems that the length of time to double might be closer to 2 years than to 18 months, but students do not have any sophisticated tools to compare the dynamically fitted models and may have answers that differ somewhat from this one.

Power Properties—Base *e*

If you invest $1000 at 5% compounded twice a year, it earns more interest than if it's compounded just once a year. Compounding monthly earns even more interest, and with computers, daily compounding is quite common. But what if compounding is continuous, every instant of time? How much would your investment be worth in 5 years with continuous compounding?

Q1 What's your guess about what the $1000 investment would grow to at 5% over 5 years if compounded continuously?

INVESTIGATE

Click on the slider's thumb to open its inspector.

1. Open a new Fathom document and pull two sliders from the shelf. Name them something like *Frequency* and *Money*. In the *Money* slider inspector, enter a formula to show the value of $1 invested at a whopping 100% per annum for one year compounded *Frequency* times every year. Keep the inspector open so you can see the value to five decimal places.

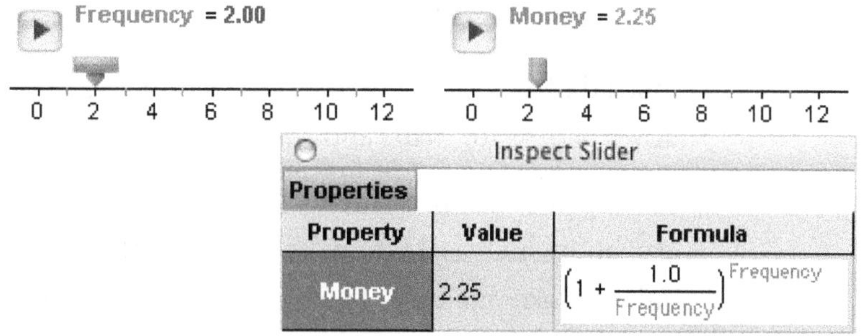

Q2 Adjust the *Frequency* slider to find how much money, to five decimal places, there will be after one year if interest is compounded with a frequency of

a. annually

b. monthly

c. daily

d. hourly

e. every minute

f. every second

How much bigger would you expect your investment of $1 to be if interest were compounded every instant (that is, continuously)?

The value 2.71828 . . . appears so often in mathematics that it has the name *e*. You have just calculated e^1 because you used the interest rate of 100%. For the 5% interest rate of your original problem, you'd use 5% instead of 100%.

2. To use different interest rates, pull down a slider called *Rate* and adjust the formula for *Money* to replace the 1.00 with *Rate*. Check to be sure you get the same value as before when *Rate* has the value 1. Most interest rates are between 0 and 0.2, so make those numbers the endpoints of the *Rate* slider.

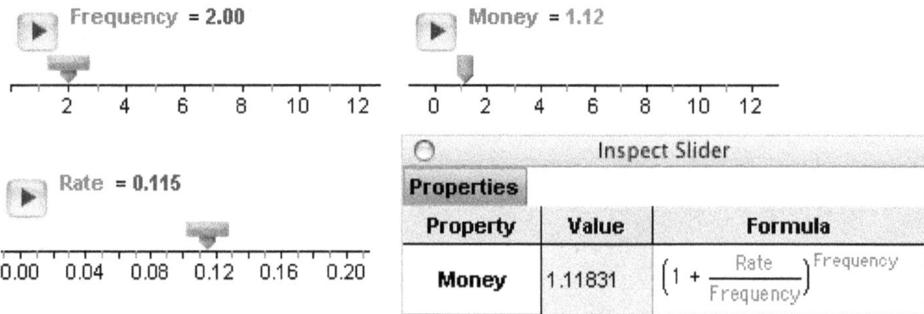

Q3 Watch the inspector of the *Money* slider as you vary the *Rate* slider between 0.0 and 0.2. How do the two values compare?

Q4 Think about investing for more than one year. How will $(e^{Rate})^{Time}$ relate to $(1 + Rate)^{Time}$? Why?

3. Instead of working with $(e^{Rate})^{Time}$, let's try to get a single exponent on *e*. Pull down sliders for *Years* and *Value*.

4. To help you see the values graphically, pull down a case table and create attributes such as *Yrs* and *Val*. Give *Yrs* the formula Years and *Val* the formula Value. Make one case. Then pull down a graph and make a scatter plot of *Val* against *Yrs*.

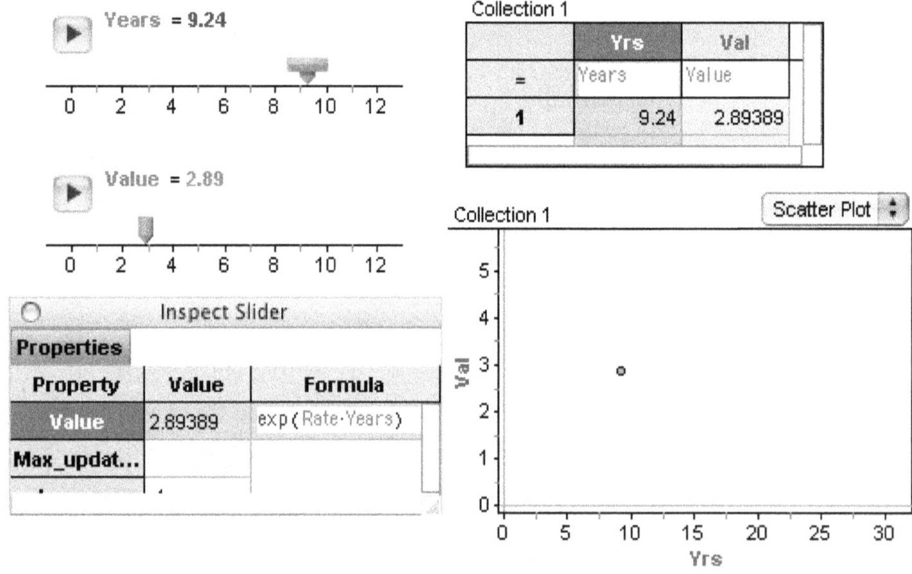

5. On the scatter plot, plot the equation Val = (exp(Rate))^Yrs.

Q5 What are good endpoints for *Val* if *Yrs* is between 0 and 30 years?

Q6 For the *Value* slider, make a formula of the form exp( ), referring only to *Rate* and *Years,* so that as the *Years* slider moves, the point (*Years, Value*) remains on the graph you just made. What formula did you use?

Q7 Equating the formula for *Value* to the formula for *Val,* what general law about powers of powers can you state?

Q8 Use your results to answer the original question: With continuous compounding at an annual rate of 5%, how much would an investment of $1000 be worth in 5 years?

EXPLORE MORE

Create a data set that records the mass of a radioactive material with a half-life of 5 years. (That is, the mass of the radioactive material will diminish by half every 5 years.) Let the initial amount be 500 g and create a case for years 0, 5, 10, . . . , 50. Graph *Mass* versus *Time.*

Half Life

	Time	Mass
1	0	500 g
2	5	250 g
3	10	125 g
4	15	62.5 g
5	20	31.25 g
6	25	15.625 g
7	30	7.8125 g
8	35	3.90625 g
9	40	1.95312 g
10	45	0.976562 g
11	50	0.488281 g

Dynamically fit an equation of the form $y = a \cdot b^x$ and $y = a \cdot e^{kx}$ to your data, where *e* is the base of natural logarithms, 2.71828. . . . Explain the significance of *a, b,* and *k* to the situation. What is the continuous rate of decrease for the mass of the radioactive material?

Objective: Students will see that continuous compounding of an investment (at rate r) for one year is modeled by the exponential equation $y = e^r$ and that this value is slightly greater than $1 + r$ for small values of r. In writing $(e^r)^t$ with a single exponent on e, students will derive a case of the exponent law $(c^a)^x = c^{ax}$.

Student Audience: Algebra 1, Algebra 2

Activity Time: 45–70 minutes

Setting: Paired/Individual Activity

Mathematics Prerequisites: Students can solve an exponential equation by systematic trial or by using technology, substitute values and expressions into formulas, and calculate amounts under compound interest with different compounding frequencies.

Fathom Prerequisites: Students can use a slider and the slider's scale, edit a plotted equation's formula, determine the value of a slider using a formula, create a collection, and add attributes and cases.

Notes: As you circulate among working pairs, watch for students who find Q6 and Q7 difficult. Ask questions such as What have you tried? How else could you combine *Rate* and *Years*? Students who finish the Explore More can work on the Extension question.

Q1 Answers will vary dramatically. Some students may think that continuous compounding yields an infinitely large value. It would be good to leave any disagreements to be settled by the exploration.

INVESTIGATE

Q2 a. When *Frequency* is 1, *Money* is 2.

b. When *Frequency* is 12, *Money* is 2.61304.

c. When *Frequency* is 365, *Money* is 2.71457.

d. When *Frequency* is $365 \times 24 = 8760$, *Money* is 2.71813.

e. When *Frequency* is $365 \times 24 \times 60 = 525{,}600$, *Money* is 2.71828.

f. When *Frequency* is $365 \times 24 \times 60 \times 60 = 31{,}536{,}000$, *Money* is 2.71828.

The amount only increased by approximately 0.00015 by compounding every second instead of every hour. Compounding continuously will produce an amount no greater than 2.71828, to five decimal places.

Q3 *Money* is a little more than $1 + Rate$. Money and rate vary directly.

Q4 $(e^{Rate})^{Time}$ will be more than $(1 + Rate)^{Time}$, because continuous compounding yields more money than annual compounding.

Q5 Good endpoints depend on the value of *Rate*. For $Rate = 0.2$, *Val* varies from 1 to 403.

Q6 $Value = \exp(Rate \cdot Years)$ or $Value = e^{Rate \cdot Years}$

Q7 $(e^{Rate})^{Years} = e^{Rate \cdot Years}$ is a special case of the exponential law of exponents: Raising a number to a power to another power is equivalent to multiplying the powers together. Symbolically, $(c^a)^x = c^{ax}$.

Q8 $1000e^{(0.05)(5)} \approx \$1284$

EXPLORE MORE

$Mass = 500(0.5)^{\frac{Time}{5}}$ and $Mass = 500\left(e^{-0.6931\frac{Time}{5}}\right)$. a is the initial amount; b is the $\frac{1}{2}$ that is being multiplied repeatedly; and k is the power to which e is raised to get b. The rate of decrease is 0.5.

EXTENSION

Write the following relationships in the form $y = ae^{kx}$.

a. $Population = 339{,}200{,}000(1.0215)^x$

b. $Pressure = 1030(0.88)^{Height\_km}$

Answers:

a. $Population = 339{,}200{,}000\left(e^{0.021272x}\right)$

b. $Pressure = 1030\left(e^{-0.1278334x}\right)$

Power Properties—Radiation

The chemical iridium 192 is used for radiation therapy in hospitals. After use, it is sent to a special storage facility where it is stored and monitored until safe. In this activity, you will study the records for one sample and predict when the sample will become safe.

Q1 Radioactivity is measured in rads. A safe level is 5 rads. The sample arrived at the storage facility measuring 80 rads. It measured only 62 rads when it was checked 28 days later. When do you expect the sample to be safe?

INVESTIGATE

Beginning on the day it arrived at the storage facility, the radioactivity has been measured every 4 weeks. The measurement data are contained in the Fathom document **Radioactive.ftm.**

Time is measured in days and *Radiation* is in rads; keep the units in mind, but don't put them in the table.

1. Open the document and create a scatter plot of the data, with the independent variable on the horizontal axis, as usual.

Q2 What is the vertical intercept? What type of graph might fit the data?

The rate of decay is less than 1, so you might set the slider to values between 0.5 and 1.

2. Use an exponential model. Create a slider. Graph an exponential equation that uses the intercept and the slider value as the base. Adjust the slider until the graph fits the points.

Q3 What equation fits the data well? What do the values mean?

Q4 Extend your graph and trace it to find when the iridium is less than 5 rads. Use a complete sentence to describe when the iridium 192 sample is safe.

The half-life of radioactive material is the time required for the amount of radiation to decrease by half. If H is the half-life, then $\frac{Time}{H}$ is the number of times the initial amount is multiplied by 0.5 in *Time* days.

3. To find the half-life of iridium 192, drag down a new slider for H. Add to the scatter plot the function radiation = 80 • 0.5 ^ (Time/H).

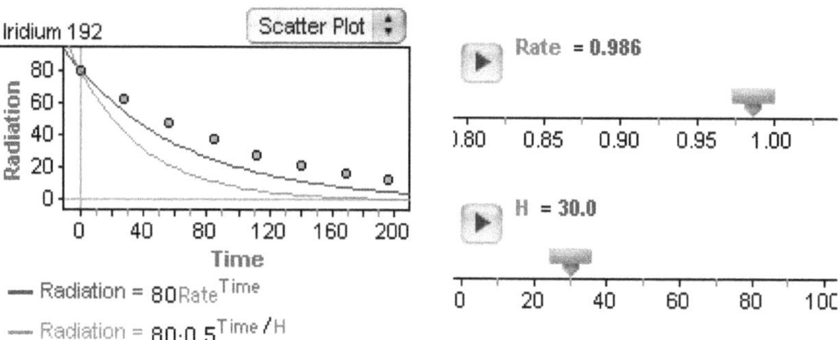

Q5 What value of H gives a graph that fits the data points?

4. To check your value of *H*, place the trace point somewhere near the middle of the graph. Record the coordinates of the point (with units). Then move the trace point *H* days earlier and record the coordinates. Finally, move to a point *H* days later than your original point and record the coordinates.

Q6 What coordinates did you record? Do these coordinates confirm that *H* is really the half-life? Why or why not?

The equation is equivalent to $Rate^H = 0.5$.

Q7 To find the half-life without using Fathom, use a calculator. Express the rate (the first slider) in terms of the value of *H* and the number 0.5.

5. Check your formula by entering it for the base slider. Now as you move the *H* slider, the two graphs should stay together.

Q8 To see a property of exponents, you can equate your two models: $80(Rate)^{Time} = 80(0.5)^{\frac{Time}{H}}$. Substitute the expression you used in Q7 for *Rate*. In words, what property does this equation illustrate?

EXPLORE MORE

Open the document **Moore.ftm** and model the data with the equation $Transistors = a \cdot 2^{\left(\frac{Years\_Since\_1971}{D}\right)}$ to calculate the doubling time *D*. Explain how you can use the property of exponents found in Q8 to determine the equation $Transistors = a \cdot b^{Years\_Since\_1971}$.

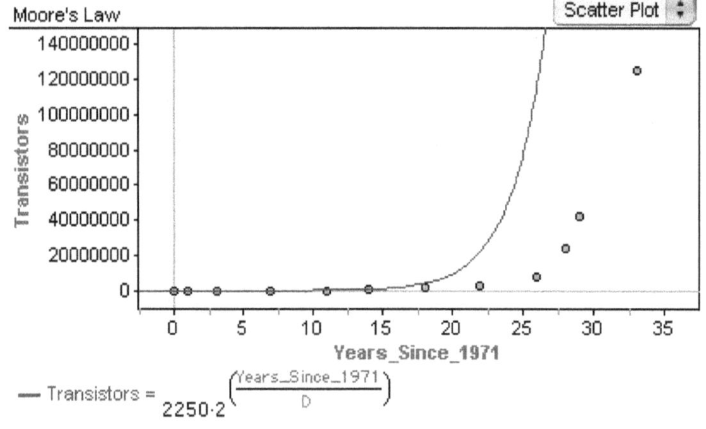

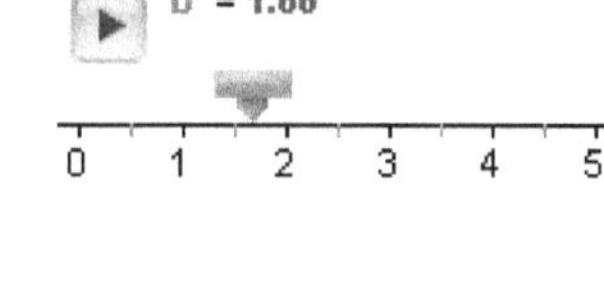

Power Properties—Radiation

Objective: Students will discover that the exponential law of exponents, $(c^a)^x = c^{(ax)}$, applies in an exponential decay model.

Student Audience: Algebra 1, Algebra 2

Activity Time: 45–70 minutes

Setting: Paired/Individual Activity (use **Radioactive.ftm**)

Mathematics Prerequisites: Students can solve a power equation and substitute values and expressions into formulas.

Fathom Prerequisites: Students can create and use sliders, enter a function with slider values, and set the value of a slider using a formula.

Q1 Answers will vary. You need not reach consensus at this time. Linear thinking leads to an estimate of about 120 days. An exponential model says it will take more than twice that time.

INVESTIGATE

1. If units are used in Fathom exponential equations, it leads to a units-incompatible message; therefore, we are not putting units in the table.

Q2 The vertical intercept is 80. Students may say that the data look parabolic or exponential. The best model for radioactive decay is a decaying exponential equation.

2. *Rate* and *Base* are descriptive names that could be used for the slider. *Rate* is used here.

Q3 Answers may vary somewhat. A good fit is $y = 80(0.9907)^{Time}$, which says the sample began with a radiation of 80 rads and loses about 1% each day.

Q4 Answers will vary. The iridium 192 becomes safe after about 300 days.

Q5 Answers will vary. *H* is approximately 74 days.

Q6 Sample answer: (96, 32.5), (22, 65), (170, 16.25). Seventy-four days earlier, the sample had twice the radiation. Seventy-four days later than the original point, the sample had half the radiation. The half-life is the amount of time it takes the radiation to decrease by half.

Q7 $Rate = 0.5^{\frac{1}{H}}$. Students might first find that $Rate^H = 0.5$.

Q8 Removing the 80 from each equation gives $\left(0.5^{\frac{1}{H}}\right)^{Time} = 0.5^{\frac{Time}{H}}$. This equation is a special case of the exponential law of exponents: Raising a number to a power to another power is equivalent to multiplying the powers together. Symbolically, $(c^a)^x = c^{(ax)}$.

EXPLORE MORE

The doubling time is about 2.09 years. You can find the base by calculating $2^{\frac{1}{2.09}} \approx 1.393$; so the alternate way to express $2250 \cdot 2^{\frac{Years\_Since\_1971}{2.09}}$ is $2250 \cdot \left(2^{\frac{1}{2.09}}\right)^{Years\_Since\_1971}$, or $2250 \cdot 1.393^{Years\_Since\_1971}$.

5

Transforming Functions

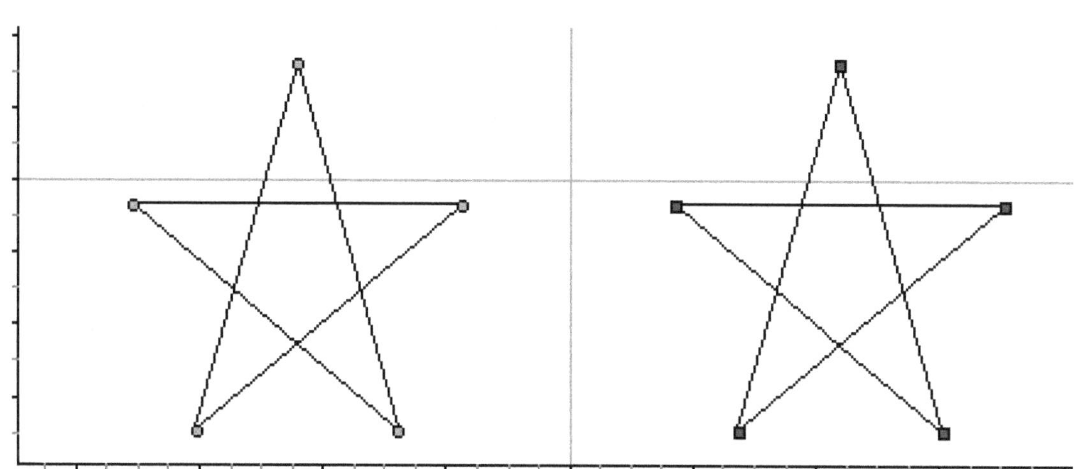

Functions—Model Rockets

Imagine launching a model rocket. The rocket's engines burn briefly at the start of flight. When the fuel is gone, the rocket continues to rise for a while before falling.

The height of the rocket at any time is described by an equation called a *function*. In this exploration, you'll learn how to use Fathom to represent and apply the machinelike qualities of functions.

INVESTIGATE

1. Open a new Fathom Document. Drag a slider from the shelf and name it Input. The slider will represent time, so enter the unit s for seconds to the right of the value.
2. Drag down another slider for height and enter the name Output. To the right of this value, enter the unit ft to indicate that the height is measured in feet.

Double-clicking on the slider thumb opens its inspector.

3. To represent the rocket's height, open the *Output* slider's inspector. Double-click in the formula box on the top line and enter the formula $\text{Output} = 40\text{ft} + 97\frac{\text{ft}}{\text{s}} \cdot \text{Input} - 16\frac{\text{ft}}{\text{s}^2} \cdot \text{Input}^2$.

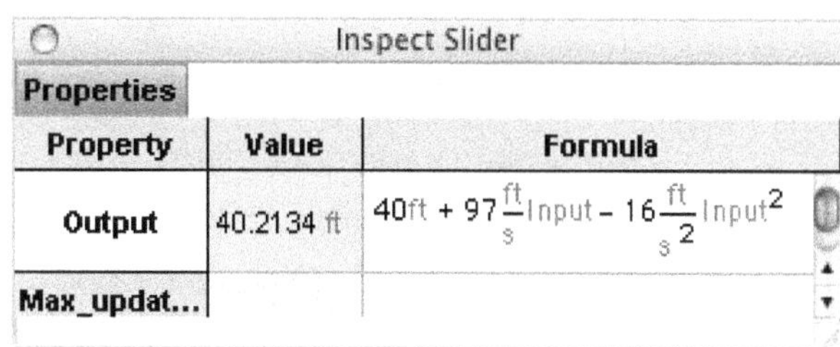

Inspect Slider

Properties

Property	Value	Formula
Output	40.2134 ft	$40\text{ft} + 97\frac{\text{ft}}{\text{s}}\text{Input} - 16\frac{\text{ft}}{\text{s}^2}\text{Input}^2$
Max_updat...		

You can think of your two sliders together as a function machine. One of the sliders depends on the value of the other, whose movement you can control.

Q1 Which of the two variables (sliders) would you call the independent variable, and which would you call the dependent variable?

The *domain* of a function is the set of input values that make sense. The *range* is the set of output values the function machine produces from input values in the domain.

Q2 Experiment with the input slider and think about the rocket. What do the input and output numbers tell you about the rocket's flight? What input values for time make sense for the flight of a rocket? This is called the *domain*.

Q3 What are the lowest and highest output values produced from the input values in the domain? Do these values all make sense for the height of the rocket? The values between these two output numbers are referred to as the *range* of the function.

4. Open the slider inspector for *Input* and set the values of *Lower_* and *Upper_* according to your domain. Then open the inspector for *Output* and set *Lower_* and *Upper_* with the values for the range.

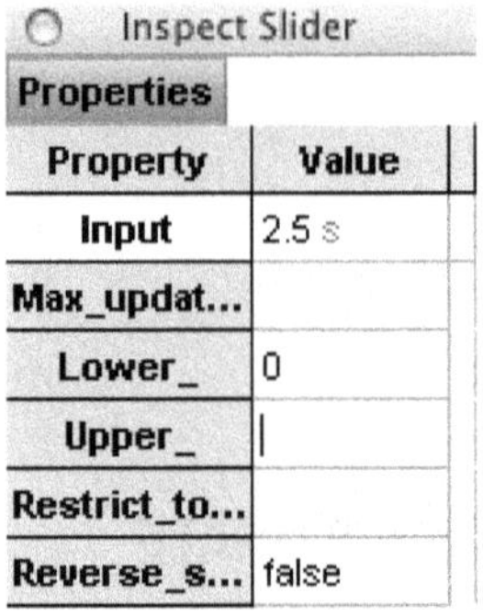

Inspect Slider

Properties

Property	Value
Input	2.5 s
Max_updat...	
Lower_	0
Upper_	l
Restrict_to...	
Reverse_s...	false

Q4 Is there a value for *Input* that produces two different values for *Output*? If so, give an example.

Q5 Are there two values for *Input* that produce the same value for *Output*? If so, give an example.

Your answer will be in the form $f(t_1) = f(t_2) = h$, where h is the height of the rocket at both time 1 and time 2.

Q6 In function notation, $f(Input) = 40\text{ ft} + 97\frac{\text{ft}}{\text{s}} \cdot Input - 16\frac{\text{ft}}{\text{s}^2} \cdot Input^2$. Use function notation to restate your answer to Q5.

Q7 What can you learn from the sliders about the highest point of the rocket's flight?

5. A graph of the function makes it easier to see which input values give the same output. To see how the function can be represented by a graph, first drag down a case table. Choose **Collection | New Cases** and add one case. Create two attributes, perhaps called *Time* and *Height*. Double-click on the collection's name and enter a name such as Rocket_Heights.

To enter the formulas, choose **Table | Show Formulas** and click on the formula fields.

6. For *Time*, create a formula that is only the name of the slider *Input*. For *Height*, use the simple formula *Output*. Check that the *Height* values match the values on the *Output* slider as you change the *Input* slider in the function machine.

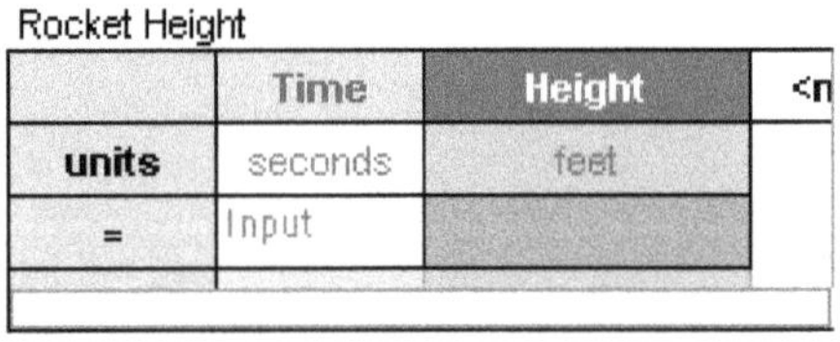

Rocket Height

	Time	Height	<n
units	seconds	feet	
=	Input		

When the cursor is over a scale, it becomes a hand with which you can adjust the scale. Or you can set *xLower* and the other limits in the graph's inspector.

7. Drag a graph from the shelf. A function graph always plots the independent variable on the horizontal axis and the dependent variable on the vertical axis. So drag your independent variable from the case table to the horizontal axis, and drag your dependent variable to the vertical axis. Adjust the scales on the resulting scatter plot according to the domain and range from Q2 and Q3.

Can you always see the plotted point as you adjust the *Input* slider? If not, check the numbers you are using on the axes' scales.

Q8 At what time does the rocket hit the ground? Write your answer in function notation.

8. In the graph, plot the function *Height* = 40 ft. Use function notation to write your answer.

Q9 What are two values for *Time* for which *Height* = 40 ft?

Q10 Consider the vocabulary *function, independent, dependent, domain, and range.* How do these words relate to each other? What will you think of to remember their definitions?

EXPLORE MORE

Drag a new graph from the shelf and change the drop-down menu from **Empty Plot** to **Function Plot.** Choose **Plot Function** from the **Graph** menu and plot a function. Explore other function graphs using various operations, including powers. For which functions are there *no* pairs of *x,* or *Input,* values having the same *y,* or *Output,* value? How does the shape of the function's graph tell you whether there are pairs of *Input* values with the same *Output* value?

Objective: Students will see function as only one output for each input but possibly more than one input with the same output. They will use the terms *independent variable, dependent variable, domain,* and *range* and see how to represent a function as a graph.

Student Audience: Algebra 1

Activity Time: 20–30 minutes

Setting: Paired/Individual Activity or Whole-Class Presentation

Mathematics Prerequisites: Students understand scatter plots and can evaluate equations.

Fathom Prerequisites: Students can create sliders with units and set scales and formulas, use the inspector to add attributes with formulas, and add cases to a new collection.

Notes: Students use Fathom sliders to create the independent and dependent variables of a function machine. The limits placed on the sliders in the inspector represent the function's domain and range. As you visit students working on Q2, you might suggest that they animate the *Input* slider. The animated *Input* slider will move at a constant speed. Ask students what they notice about the speed of the *Output* slider. Students may also find it interesting to animate the *Input* slider after they have created the graph in step 7. Though the rocket will also travel in a parabolic path, what students see on the animated graph is not the path of the rocket but a function describing the height of the rocket over time.

For a Presentation: A whole-class presentation could introduce or review function vocabulary. Questions Q2–Q5 can lead to good discussion. Animating the *Input* slider can lead to further discussion. Asking several students to share their thinking on Q10 can benefit both the students who are speaking and those listening.

INVESTIGATE

Q1 The independent variable is *Input,* and the dependent variable is *Output.*

Q2 Input shows time in flight; output gives height of rocket. The domain should cover the values from 0 s to about 6.5 s.

Q3 The range should cover values from 0 ft to about 190 ft. Positive values make sense. Again, students will differ somewhat on the endpoints.

Q4 There is no *Input* value that gives more than one *Output.* If there were an *Input* value that gave more than one *Output,* then the output would not completely depend on the input, and this would not be a function.

Q5 Many answers are possible; for example, 1 s and 5.07 s both give about 121 ft.

Q6 Answers will depend on the answer to Q5. For the sample, $f(1) = f(5.07) = 121$ ft.

Q7 The highest the rocket will go is about 187 ft at about 3 s.

7. Graph windows should be at least as big as *xLower* = 0 s, *xUpper* = 6.45 s, *yLower* = 0 s, and *yUpper* = 1.87 s.

Q8 The rocket hits the ground at about 6.45 s; $f(6.45) = 0$.

8\.

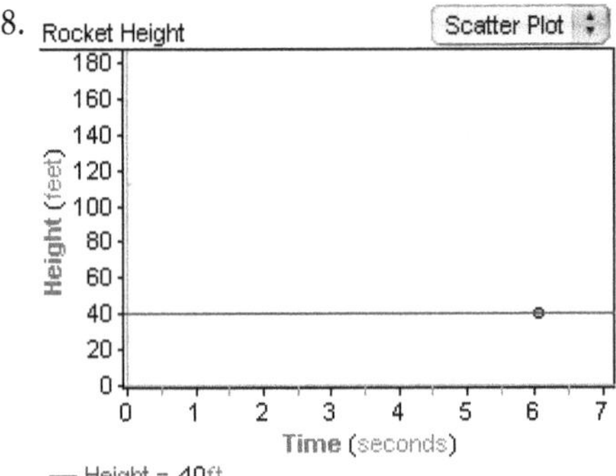

Q9 The rocket will have a height of 40 ft at both 0 s and 6.06 s; $f(0) = f(6.06) = 40$.

Q10 Answers will vary.

EXPLORE MORE

Answers will vary. Common functions for which there are no pairs of *Input* values having the same *Output* value are linear and exponential functions and power functions with odd powers (for example, $f(x) = x^3$). Sample descriptions of the graph: If the graph doesn't "turn around" (have any relative maxima or minima), then a function with an unbroken graph won't take two different output values to the same input value. No horizontal line crosses the graph more than once.

Data Transformation—Quiz Scores

A teacher gives a quiz on which students do not score very well. The top score on the quiz is 72 out of 100 points. The teacher doesn't want this one quiz to bring down students' grades for the entire year.

Q1 What would you recommend the teacher do?

INVESTIGATE

1. The scores are in **Quiz Scores.ftm.** Open the document and make a dot plot of *Scores.*
2. One method of adjusting scores is to add points. Drag down a slider and call it something like *Shift.* In the slider's inspector, set *Restrict_to_multiples_of* to 1 to be sure the adjusted scores will be integers. Set the scale to include negative values.
3. Add an *Adjusted_Scores* attribute to the table. Enter the formula Scores + Shift. Drag *Adjusted_Scores* to the horizontal axis of the graph but don't drop it. A plus sign will appear below the axis on the left. Drop *Adjusted_Scores* on the plus sign.

You'll see both the original dot plot and the dot plot for *Adjusted_Scores.* Data that have a dot plot with a tail of spread-out values on the left, like these, are described as *skewed left.*

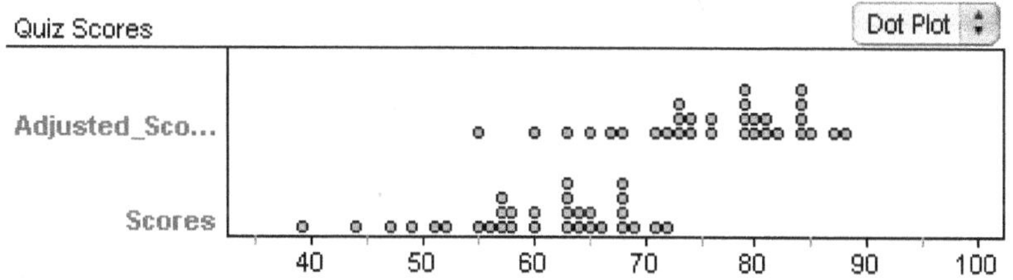

Q2 How does the dot plot for *Adjusted_Scores* change as *Shift* changes? Explain any patterns you see.

Q3 If you shift the largest score to 100, what will the smallest score become? Why?

4. In the dot plot, plot the values of the median.

With the graph selected, choose **Graph | Plot Value** and type median(Scores). Repeat for median(Adjusted_Scores). To avoid typing, double-click on the function (for median, **Functions | Statistical | One attribute | Median**), then double-click on the attribute name.

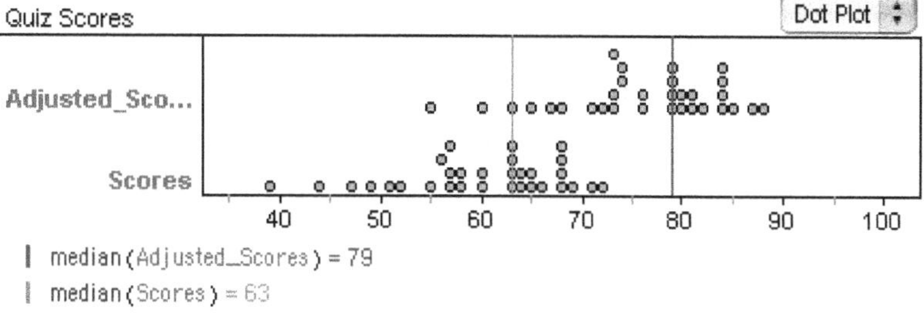

Q4 How does the median of the scores change as *Shift* changes? Why?

Q5 How does the shape of the data set change as *Shift* changes? Why?

The teacher thinks it might be unfair to add the same amount to the lower scores as she adds to the higher scores. Multiplication is another option.

5. Drag down a slider for *Stretch.* Set the scale for *Stretch* so that the top score doesn't go over 100. Change the formula for *Adjusted_Scores* to be a product by *Stretch,* rather than a sum.

Q6 How does the dot plot change as *Stretch* changes? Explain the patterns you see. Describe what happens when *Stretch* is between 0 and 1 and when *Stretch* is negative.

Q7 How does the shape of the data set change as *Stretch* changes? Explain.

Q8 If you multiply to make the largest score 100, what will the smallest score become? What is the new median?

Q9 What do you think is the fairest way to adjust the scores so that no score is greater than 100 or less than 0?

EXPLORE MORE

1. The teacher would like to find a way to combine the transformations so that the median score is 75 and the high score is 100. Combine the shift with the stretch to find the values needed for this adjustment. Are all students likely to be happy with this adjustment? Find a way that you think is fair to adjust the score so that the high score is 100.

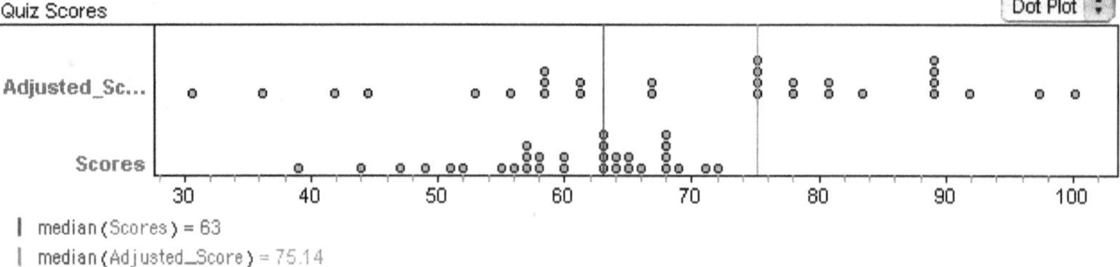

Use sqrt( ) for the square root function.

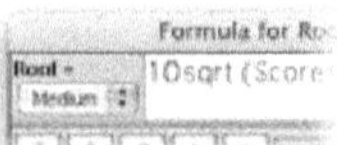

2. When *Scores* are between 0 and 100, some teachers use the formula $10\sqrt{Scores}$. Compare the graph of this transformation with the original scores. How are the maximum, the median, and the shape affected by this? Find your own new transformation formula and explain what it does.

Data Transformation—Quiz Scores

Objective: Students will see how the position and spread of a data set are changed by shifts and stretches, even though the shape of the data set remains the same. They will review the median of a data set.

Student Audience: Algebra 1, Algebra 2

Activity Time: 25–35 minutes

Setting: Paired/Individual Activity (use **Quiz Scores.ftm**)

Mathematics Prerequisites: Students understand dot plots, medians, spread, and shape.

Fathom Prerequisites: Students can create and name sliders and set the scales, add attributes and give them formulas, create a dot plot and plot values, and add a second graph to an existing one.

Notes: As you visit student pairs and talk with them about what they are noticing, you might introduce the word *translate* in place of *shift.* As you consider students to present their work to the class, look for different answers to Q9 and for students who can share results from the Explore More questions. As you visit groups, listen for such vocabulary as *positive, negative, horizontally, reflected,* and *decrease.* Use prompts to encourage full, precise descriptions: Can you say it another way? What else do you see? How else are they different? Alike? What if the shift is negative? Look for opportunities to use the terms *translation* and *dilation.*

Q1 Answers will vary. Some students may think that no adjustment should be made, but most will suggest adding up to 28 points to everyone's score.

INVESTIGATE

Q2 The entire plot shifts to the right when *Shift* is positive, because the numbers are getting larger, and larger numbers are to the right on the standard number line. If *Shift* has a negative value, the points are translated to the left.

Q3 The smallest value shifts from 39 to 39 + 28, or 67.

Q4 The median shifts exactly with the data, because the middle remains the middle.

Q5 The shape doesn't change as the data are shifted.

Q6 The plot stretches or shrinks horizontally, depending on whether *Stretch* is greater than or less than 1. The range is multiplied by *Stretch*, because (*Stretch*)(max) − (*Stretch*)(min) = (*Stretch*)(max − min). Negative values cause the data to flip to the other side of zero as a mirror image.

Q7 Although the spread of the data is changed, the shape (skewed left) is not, except to be reflected (skewed right) when *Stretch* is negative. The data set changes as if it were plotted on a sheet of rubber that is nailed at zero. You can stretch it or squish it or flip it, but the shape is still the same, because the treatment is uniform.

Q8 Multiplying scores by 1.39, to move 72 to about 100, changes 39 to about 54.2. The median is moved from 63 to 87.5 points.

Q9 Answers will vary.

EXPLORE MORE

1. To move the scores to these values, you must multiply by about 2.78 and shift −100. This is not good for students with a score below 57, because their score actually goes down with this formula. Students will have many of their own ideas.
2. The square root function raises all the scores and causes a shape change, so these data become less skewed. This transformation is not a stretch, because it increases smaller values more than it does larger ones. Students may wish to combine this effect with the shifting and stretching they played with in the activity.

Transformations—Animation

Graphic artists who create animated movies or video games need to make objects move around on a screen. In this activity, you'll see the mathematical basis of computerized animation.

Q1 How do you think computers make screen objects move around and stretch and shrink?

INVESTIGATE

1. Open the document **Star.ftm.** This data set contains the coordinates of the vertices of a five-pointed star. You'll be moving a star like this around on the screen.

In the upper right corner of the graph window, click on the drop-down menu and choose **Line Scatter Plot.**

2. To see what the basic star looks like, pull down a new graph. Drag the x-attribute to the horizontal axis and the y-attribute to the vertical axis. Change the plot type to a line scatter plot. The points will be connected in the order they appear in the table.

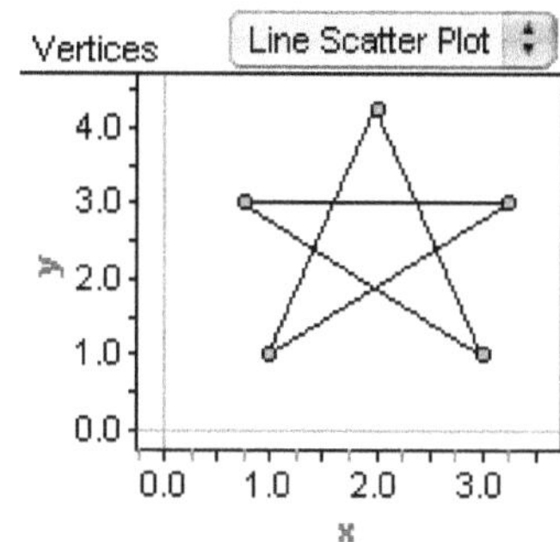

3. Now you are going to create the animation controllers. Drag down a slider and name it something like *H_Shift* to represent the amount by which the points will be shifted horizontally.

Show the formulas in the table. Enter x + H_Shift as the formula for this new coordinate.

4. The attributes x and y hold the original shape of the figure. You will create new variables for the animation. In the table, create an attribute for the new x-coordinate, perhaps called *x_New.* This variable should involve both the original data and the value from the slider.

5. To see both graphs, add the *x_New* attribute to your graph's horizontal axis by dragging it down and releasing it on the plus sign.

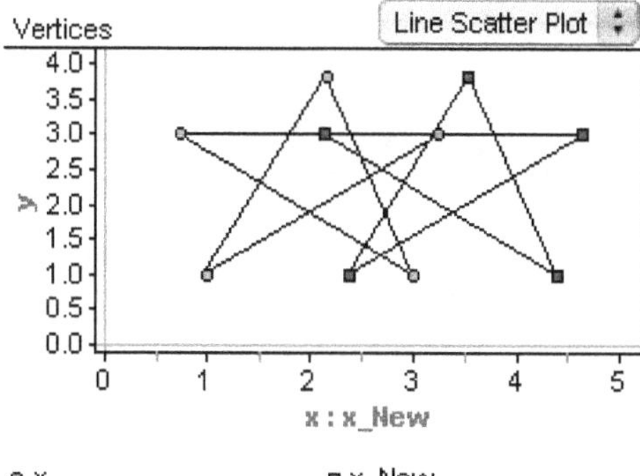

Q2 How does this new figure differ from the original star?

Q3 What happens as you slide the *H_Shift* values? Explain what you see.

Q4 Check that the scales on the slider and the graph include negative numbers. What happens to the star if *H_Shift* is negative? Explain what you see.

Q5 If you drag vertices of the original star to change its shape, what happens to the shifted star?

6. To add vertical animation, drag another slider from the shelf for vertical shifts, perhaps called *V_Shift*. Create a new variable, *y_New*, for the new vertical coordinate and give it an appropriate formula.

Be sure the graph is selected. Then choose **Graph | Remove X Attribute: x_New.**

7. To see this transformation, you need to swap attributes. Remove *x_New* and add *y_New* to your graph's vertical axis.

Q6 What happens as you slide the *V_Shift* values? Include negative numbers.

You've seen how graphic arts software can move an object around on a screen by adding to the coordinates. What happens if the software multiplies the coordinates by numbers?

8. For the multiplying number, pull down another slider for a variable called *H_Str*. Change the formula for *x_New* to x • H_Str + H_Shift.

9. To make a multiplier for the attribute *y_New*, pull down yet another slider (*V_Str*) and enter the formula y_New = y • V_Str + V_Shift.

10. To view all the changes, duplicate the graph. Drag *x_New* to the horizontal axis and *y_New* to the vertical axis. Change the graph to a line scatter plot.

Q7 What happens to the shape of the star as you slide the *H_Str* and *V_Str* values? Include negative values for the sliders. Explain.

Q8 To animate sliders, click on the arrows next to their names. Adjust the scales as needed to see dramatic changes. What scales give the best effects?

EXPLORE MORE

1. Adjust scales and sliders to make the star start small at the bottom left of the graph and grow as it moves to the top right of the graph.

2. Animate a figure of your own design. For *x* and *y*, enter the coordinates of the vertices in the order you want them to be connected.

3. Rotate by creating a slider called *Angle*. Enter the formula x_New = x • cos(Angle) – y • sin(Angle) and y_New = x • sin(Angle) + y • cos(Angle). What happens as *Angle* changes?

Transformations—Animation

Objective: Students will explore two-directional translation and dilation of line plots.

Student Audience: Algebra 1, Algebra 2

Activity Time: 45–60 minutes

Setting: Paired/Individual Activity (use **Star.ftm**)

Optional Document: Moving Star.ftm

Mathematics Prerequisites: Students can identify coordinates of a given point and plot a point given its coordinates on scaled rectangular coordinate axes. They understand that negative numbers are usually to the left of or below zero on a number line and that the farther from zero a negative number is, the lower or farther to the left it is pictured on a number line.

Fathom Prerequisites: Students can create a new attribute and enter a closed-form formula for an attribute, show formulas in a table, adjust scales on graphs, add a second attribute to one of the axes of a graph, and use a slider.

Fathom Skills: Students learn how to remove an attribute from an axis of a graph and to animate a slider.

Notes: Before students begin, you may wish to animate the slider in the document **Moving Star.ftm** to spark some interest in the activity. As you listen in on students' descriptions and talk with them about what they are doing and seeing, listen for and model accurate vocabulary, such as *increasing* and *decreasing, positive* and *negative, horizontal* and *vertical,* and *reflection.* Introduce the term *translation* as students work on Q6 or *dilation* as they work on Q7. To encourage complete descriptions, ask, What if the slider is less than one? Equal to zero? Negative? How are the coordinates changing?

Q1 Students may have many quite varied ideas about how computer animation works.

INVESTIGATE

Q2 If the slider is still set at its initial value of 5, the new star will be 5 units to the right of the first one.

Q3 The star shifts horizontally to the right of the original star for positive values and to the left for negative values.

Q4 As the slider moves to the left beyond zero and the *H_Shift* values become negative, the new star moves to the left of the original star. The horizontal coordinates become less than those of the original star.

Q5 The shifted shape changes in the same way the regular shape does.

Q6 As the *V_Shift* values increase, the star moves up; as they decrease, it moves down. If the values are negative, the moving star is below the original. You might use the term *translation* in place of the word *shift.*

Q7 The value for *H_Str* is the amount by which the star stretches horizontally. If the value is less than 1, then the stretch is a shrink. If it's negative, the star is reflected horizontally. Similarly, the value for *V_Str* is the amount by which the star stretches vertically. If the value for *V_Str* is negative, the star is reflected vertically, so that it points downward. You might introduce the term *dilating* to include stretching and shrinking.

Q8 Answers will vary. The aim is to have students see the effects of animation.

EXPLORE MORE

1. One way is to have the shift sliders run from -30 to -5 and the dilation sliders from 1 to 10.
2. Even the most reluctant student might become engaged in animating a letter that's one of their initials or that of a friend.
3. The star rotates about the point (0, 0).

EXTENSION

What is the effect of changing coordinates by subtracting and dividing, as was already done with adding and multiplying?

Answer: When applied to coordinates, effects of subtraction and division are opposite those of addition and multiplication. Thinking about subtraction and division now can lay the groundwork for changing formulas to accomplish transformations. If the initial figure is given by $y = f(x)$, then the transformation (using the slider variables above) is given by $\frac{y - V\_Shift}{V\_Str} = f\left(\frac{x - H\_Shift}{H\_Str}\right)$.

The formula for the transformation will then be $y = V\_Shift + V\_Str \cdot f\left(\frac{x - H\_Shift}{H\_Str}\right)$.

Line Transformations—Elevator

You're working for a company that will install elevators in a new 22-floor building. One elevator will make nonstop trips to the observation deck on the 22nd floor. Your supervisor wants you to create a function indicating where that elevator is at any time on its trip upward.

The architect says that the floors are 13 feet apart. Your supervisor tells you that the elevator will travel at 1.1 seconds per floor.

Q1 Make a conjecture: What function expression will give the elevator's height at any time?

INVESTIGATE

The value of *caseIndex* is the number of cases in the table. To add cases, choose **Collection | New Cases.**

Elevator

	Floor
units	
=	caseIndex – 1
1	0
2	1
3	2

1. Open a new Fathom document. Drag a case table from the shelf and set up attributes for *Floor* and *Height.* Give the case table a name. From the **Table** menu, choose **Show Units** and **Show Formulas.** Be sure to include appropriate units for *Height.*
2. You could enter the numbers from 0 to 22 for the *Floor* attribute, or you could enter the formula caseIndex – 1. Floor 0 is the basement. Add 23 new cases to your table.
3. To impress your supervisor, you will use units in your formula. Set up the unit fl for floors. (Confirm that you want to establish a new unit.)

Using the given information, take your best guess at a formula for the *Height* attribute that will list the heights of the floors above the basement.

4. Enter your formula in the table. Remember to include units in your formula values.

Q2 What formula did you enter for *Height*? What are the units?

5. To see if the formula is right, make a scatter plot of *Floor* and *Height.*

The supervisor happens to see your graph and says that the function should give the height of the elevator above ground rather than above the basement. The basement is 8 feet below ground. To allow for further changes in instructions, you decide to make the starting height flexible.

6. Drag down a slider and name it Base_Height. Make sure to give it an appropriate unit. Change the formula for *Height* so that it gives the distance above *Base_Height* rather than the distance above zero. Adjust the scales on the slider and the graph to allow negative values for *Base_Height.*

Q3 What formula are you using for *Height* now?

Q4 What happens to the line of points as you drag the *Base_Height* slider to different values? Why?

Q5 If the basement is at −8 ft, how can you tell from your graph the height of the 22nd floor?

7. Before you let your supervisor see this graph, you want to express the height of the elevator as a function of time. Create a new attribute called *Time* and enter seconds for the units. Create a formula for the *Time* attribute in terms of *Floor.*

Q6 What formula are you using for *Time*? Be sure to include units.

8. Change the graph to plot *Time* and *Height.* Leaving *Base_Height* as −8 ft, click on the point on the graph that represents floor 5. Check that the fifth floor is highlighted in the table.

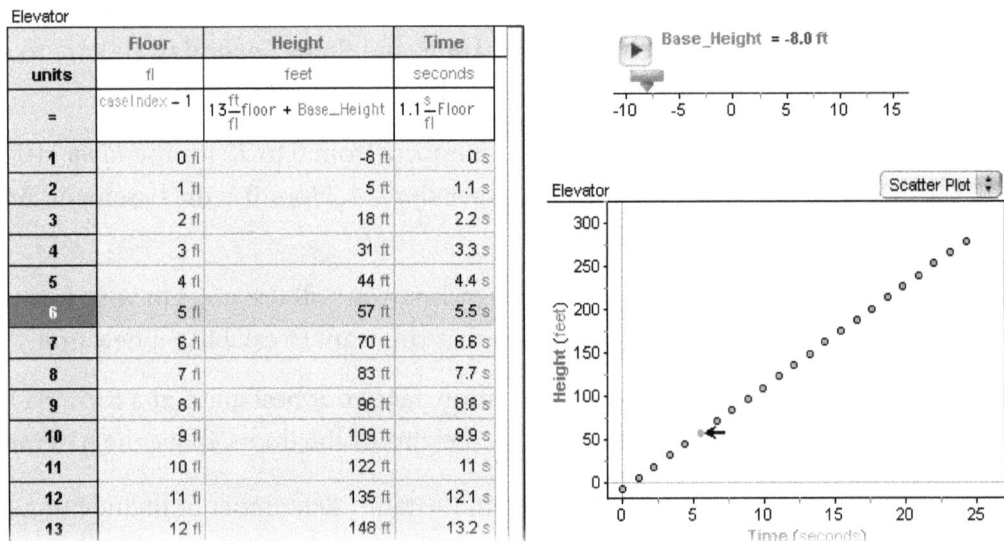

Elevator

	Floor	Height	Time
units	fl	feet	seconds
=	caseIndex − 1	$13\frac{ft}{fl}$floor + Base_Height	$1.1\frac{s}{fl}$Floor
1	0 fl	-8 ft	0 s
2	1 fl	5 ft	1.1 s
3	2 fl	18 ft	2.2 s
4	3 fl	31 ft	3.3 s
5	4 fl	44 ft	4.4 s
6	5 fl	57 ft	5.5 s
7	6 fl	70 ft	6.6 s
8	7 fl	83 ft	7.7 s
9	8 fl	96 ft	8.8 s
10	9 fl	109 ft	9.9 s
11	10 fl	122 ft	11 s
12	11 fl	135 ft	12.1 s
13	12 fl	148 ft	13.2 s

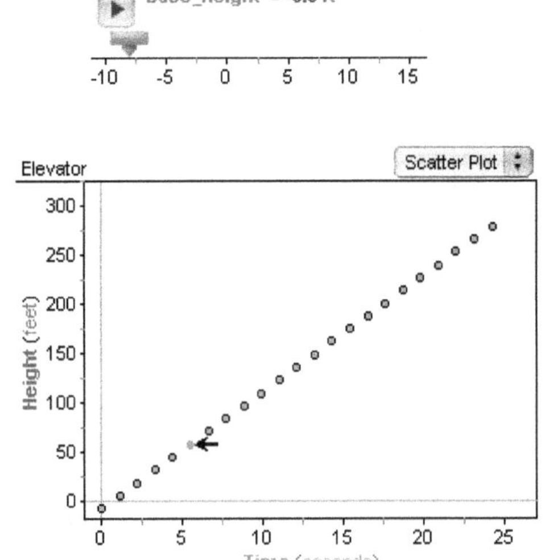

Q7 What time and height does floor 5 have if floor 0 is changed to +4 ft?

To further impress your supervisor, you decide to have Fathom draw a line through the data points.

9. Set *Base_Height* to 0. Choose **Plot Function** and experiment with equations until you've drawn a line through the data points.

Q8 What function expression did you use?

Having learned from your supervisor's earlier comment, you decide to make the first time flexible, corresponding to a floor, but not necessarily corresponding to floor 0.

10. From the shelf, drag a slider for *Base_Time,* measured in seconds, and change the formula for *Time* to take *Base_Time* into account.

Q9 Sure enough, your supervisor decides that *Time* = 0 should correspond to *Floor* = 1. In that case, what time corresponds to the point representing the 22nd floor? What time corresponds to the basement point?

Q10 In general, what happens to the line of points when you drag *Base_Time* values? Why?

11. Continue your work by making similar changes to the function. Plot an improved height function to go through the dots when *Base_Height* = 0 and *Base_Time* = 5. You might think of the existing solid line as 5 s to the left.

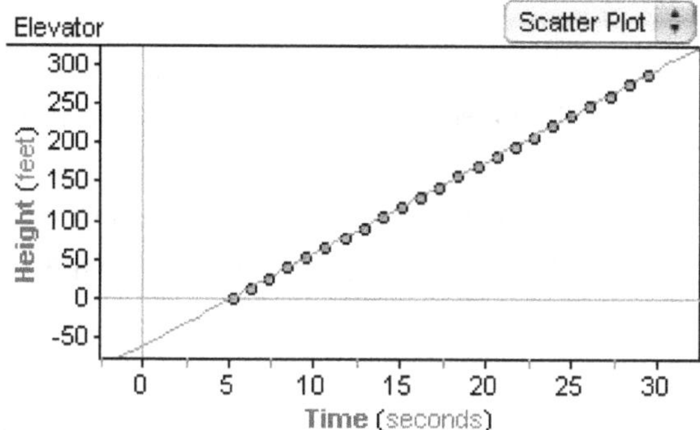

Q11 What function expression did you use?

Q12 What function expression has a graph that is a line through the dots for any values of *Base_Height* and *Base_Time*? Edit your function and make sure it follows the points as you drag the sliders.

EXPLORE MORE

1. The engineers decide to slow down the elevator because too many people may get motion sickness. Introduce a slider for the rate, measured in seconds per floor. Change the *Time* attribute accordingly, then explain what happens to the graph as you slide the new slider's values.
2. Use the graphs to determine how long it will take the elevator, traveling at different rates, to reach the observation deck, considered to have *Height* = 286 ft. (You can use *Base_Height* = 0 ft and *Base_Time* = 0 s.)
3. Suppose you know that *Base_Height* = −8 ft, that the elevator is traveling at 2 s/fl, and that the elevator passes the 4th floor after 7 s. How can you find the time at which the elevator reaches the observation deck at floor 22? Explain why you made each change that you did.

Objective: Students will learn about vertical and horizontal translations of lines and relate transformations of data to transformations of functions. Explore More 1 uses dilation to change the steepness of a line.

Student Audience: Algebra 1

Activity Time: 30–45 minutes

Setting: Paired/Individual Activity

Mathematics Prerequisites: Students can write a linear expression for motion, given speed and initial position. They understand how a graph represents a function expression and how a straight-line graph can represent motion at a constant speed.

Fathom Prerequisites: Students can set up attributes, make new cases, and work with formulas and units in a case table; use movable lines and sliders; and plot and edit functions.

Fathom Skills: Students create their own units.

Notes: This exploration can help students understand why they subtract from x in an equation to shift the graph to the right, whereas they add to x to shift a data point to the right. Q4 and Q10 focus students' attention on the direction of the shift of a line—up when a positive number is added to the entire expression and to the right when a positive number is subtracted from x alone. For a line, any vertical translation is equivalent to a horizontal translation. The amount the line shifts vertically is the product of its slope with the amount of horizontal shift. You might show an example: Graph $y = 3 + 2x$; subtract 3 from the x to get $y = 3 + 2(x - 3)$ or $y = -3 + 2x$; graph this line to see it has shifted down by 6 and to the right by 3. The activity Parabola Transformations—Handshakes gives students an opportunity to think about this further.

The terms *slant* and *slope* describe the graph. The term *rate of change* refers to the function. The term *rate* refers to the context (here, the speed of the elevator). Help students make the connection between these equivalent terms and use them correctly.

Q1 Answers will vary.

INVESTIGATE

Q2 $Height = 13\frac{\text{ft}}{\text{fl}} \cdot Floor$; units are feet.

Q3 $Height = 13\frac{\text{ft}}{\text{fl}} \cdot Floor + Base\_Height$

Q4 The graph shifts (translates) vertically, because *Height* is on the vertical axis and its values are changing.

Q5 Clicking on the graph's rightmost data point highlights a row of the table that shows the height is 278 ft. Or simply point to that point on the graph and read the coordinates from the status bar.

Q6 $Time = 1.1\frac{\text{s}}{\text{fl}} \cdot Floor$

Q7 $Time = 5.5$ s; $Height = 69$ ft

Q8 $Height = 11.82\frac{\text{ft}}{\text{s}} \cdot Time$ or $Height = \frac{13\,\text{ft/fl}}{1.1\,\text{s/fl}} \cdot Time$

Q9 *Time* is about 23 s for the 22nd floor and -1.1 s for the basement.

Q10 The graph shifts (translates) horizontally, because *Time* is on the horizontal axis and its values are changing.

Q11 $Height = 11.82\frac{\text{ft}}{\text{s}}(Time - 5\text{ s})$ or
$Height = \frac{13\,\text{ft/fl}}{1.1\,\text{s/fl}}(Time - 5\text{ s})$

Q12 $Height = 11.82\frac{\text{ft}}{\text{s}}(Time - Base\_Time) + Base\_Height$
or
$Height = \frac{13\,\text{ft/fl}}{1.1\,\text{s/fl}}(Time - Base\_Time) + Base\_Height$

EXPLORE MORE

1. Assuming the slider is called *Rate* and is set up to have units $\frac{\text{s}}{\text{fl}}$, the formula for *Time* becomes $Rate \cdot Floor + Base\_Time$. The graph through these points becomes $Height = \frac{13}{Rate}\frac{\text{ft}}{\text{fl}}(Time - Base\_Time)$. The graph changes slant, or slope, as the rate of change varies. As the rate increases, the time increases (the elevator travels more slowly), so the slope of the graph decreases.
2. Students can use a variety of approaches: They can find the first coordinate of the rightmost data point (whose second coordinate is always 286). They can eyeball the graph or, better, plot the function $Height = 286$ ft to see where the graph crosses this line. In general, the height will be 286 ft at time $22 \cdot Rate$ s.
3. Set the *Rate* slider to $2\,\frac{\text{s}}{\text{fl}}$ and adjust the *Base_Time* slider to make the fifth point (floor 4) have $Time = 7$ s. (It is easy to see on the table; when *Time* for floor 4 is 7 s, *Base_Time* is -1 s.) Then look in the table to find that at the observation deck, time is 43 s.

Parabola Transformations—Handshakes

Before each student council meeting, everyone in the room shakes hands with everyone else. They do the same at the end of the meeting. Your goal is to predict how many handshakes will take place at the next meeting.

Q1 How many handshakes do you guess will take place if 20 people attend the meeting?

INVESTIGATE

1. In a new Fathom document, drag down a new case table. Set up attributes for *People* and *Handshakes* and enter data points for the number of people from 2 through 7. With 2 people, there would be 1 handshake at the start and 1 at the end of the meeting, so record 2 in the handshake column. For 3 people, there would be 3 handshakes at the beginning and 3 at the end. For 4 people, the total is $2 \cdot 6$, or 12. Continue to calculate, draw pictures, and look for patterns until you have handshake values up to 7.

2. To look for a general formula, make a scatter plot of *Handshakes* against *People*. The data points look as though they might lie on a parabola. The parent function for a parabola would have the equation $Handshakes = People^2$. Graph that function.

Q2 How might you move the graph to make it fit the data well?

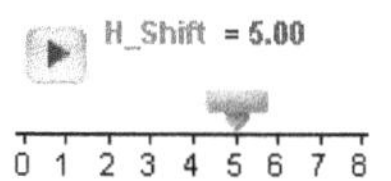

3. To shift the parabola, drag down a slider for horizontal shifting.

4. Modify the parabola's equation to shift it horizontally by the amount *H_Shift*. It may help to realize that the height of a point on the shifted parabola will be the same as its height on the original parabola. Be sure the parabola shifts horizontally as you slide *H_Shift*.

As *H_Shift* increases, the parabola shifts to the right.

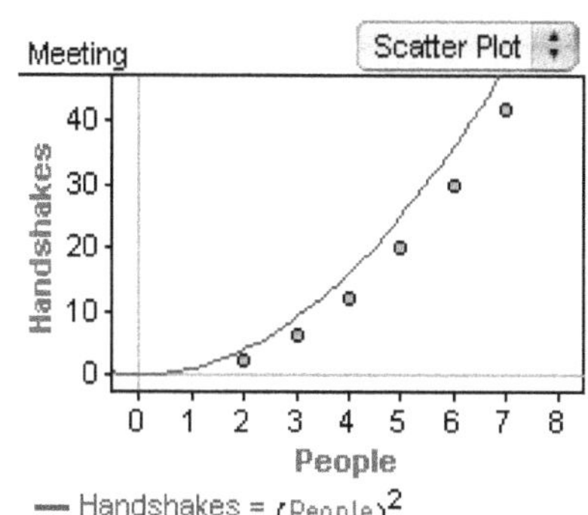

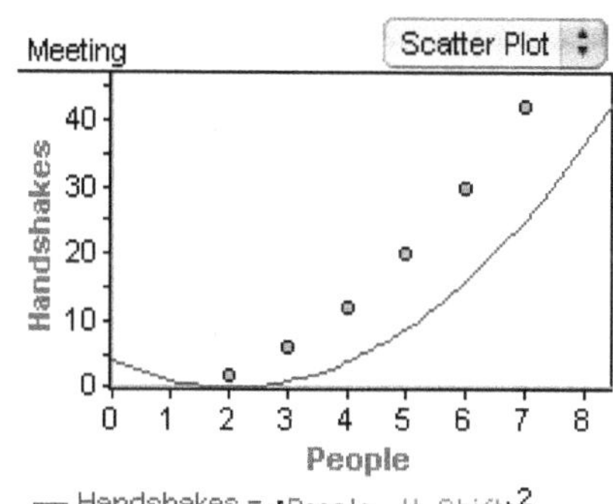

Q3 What formula did you use for the shifted parabola?

Q4 What value of *H_Shift* puts the vertex of the parabola directly below the leftmost data point?

5. You can match the data fairly well with just *H_Shift*. Look for another good match that shifts the parabola vertically, as well. Create a slider *V_Shift*.

Q5 What formula did you use for this shifted parabola?

Q6 What value of *V_Shift* will now put the vertex of the parabola at the leftmost data point?

To see small differences between the function graph and the dot plot, you will need to enlarge your graph and expand the vertical scale.

Q7 What values of *H_Shift* and *V_Shift* can you find to make the parabola fit the data points? Try small negative values for *V_Shift*. The vertex of the parabola will not be at 2 *People*.

Q8 According to the graph and equation you found in Q7, how many handshakes will take place if 10 people attend the meeting? How many handshakes at a meeting of 20 people?

EXPLORE MORE

Select the graph and go to **Graph | Show Squares.** Use the squares to improve the fit of the parabola to the data points. Because this is exact data (if the six values you entered are correct), the model should be exact, and the sum of the squares should be zero.

Parabola Transformations—Handshakes

Activity Notes

Objective: Students will learn to modify a quadratic equation in order to translate its graph.

Student Audience: Algebra 1, Algebra 2

Activity Time: 30–45 minutes

Setting: Paired/Individual Activity or Whole-Class Presentation

Mathematics Prerequisites: Students understand the relationship between a graph and its equation.

Fathom Prerequisites: Students can set up a case table with attributes and enter data, create sliders and give them names, create a scatter plot, and plot a function using a formula that refers to sliders.

Fathom Skills: Students who complete the Explore More will learn how to show squares to help with fitting a curve to data.

Notes: Students begin by creating data for the classic handshake problem. They experiment with sliders to see how changing the equation of the parent parabola can move it horizontally and vertically. The key ideas to emphasize as students talk about their work are that *H_Shift* is subtracted from *People* before the term is squared and *V_Shift* is added to the right side of the equation. These mathematical data can be modeled exactly by the formula $People(People - 1)$, which is equivalent to $\left(People - \frac{1}{2}\right)^2 - \frac{1}{4}$. For students who see $People(People - 1)$ as a way to model the data, encourage them to show that it is equivalent to the formula found by graphing the data and finding the best fit. Or use the extension once students have found the best-fit parabola.

For a Presentation: You could change the context to whatever gathering engages your students. If you have students who may not touch other students for religious or medical reasons, you can change handshaking to vocalizing individual greetings.

Q1 Guesses will vary widely. A common one may be $20 \cdot 20$, or 400. Leave the question open at this time.

INVESTIGATE

1. Students may have difficulty finding the numbers of handshakes. As needed, suggest that they draw diagrams or actually do the handshaking among themselves.

People	Handshakes	People	Handshakes
2	2	5	20
3	6	6	30
4	12	7	42

Q2 Answers will vary. All that is really required is shifting the parabola horizontally and vertically. Introduce the word *translation* in place of *shift.*

4. Some students will want to add *H_Shift* to *People,* rather than subtract it. Encourage them to think about the second sentence: "The height of a point on the shifted parabola will be the same as its height on the original parabola." You might add: As the parabola shifts horizontally to the right, the y-value of a point on the parabola does not change; to find a y-value, you can think of going back to the unshifted parabola by subtracting from x.

Q3 $Handshakes = (People - H\_Shift)^2$

Q4 $H\_Shift = 2$

5. $H\_Shift = 0.53$, with no *V_Shift,* matches the points well.

Q5 $Handshakes = (People - H\_Shift)^2 + V\_Shift$

Q6 $V\_Shift = 2$

Q7 Answers will vary. The best fit is at $H\_Shift = 0.5$ and $V\_Shift = -0.25$.

Q8 Answers will vary according to the equation in Q7. The theoretical number for 10 is 90; for 20, it is 380.

EXPLORE MORE

Squares exaggerate differences between the curve and the data points, allowing students to see near misses.

EXTENSION

Once students have the formula *Handshakes* = $(People - 0.5)^2 - 0.25$, ask them to prove that this is the same as *People*(*People* − 1). Challenge them to explain, perhaps using diagrams, why *People*(*People* − 1) makes sense as a formula for determining the number of handshakes.

Answer: Each person (*People*) shakes hands with everyone except himself (*People* − 1). The number of handshakes before the meeting is half that number, because we have counted a handshake between 2 people twice. When the same number of handshakes is given at the end of the meeting, the total is *People*(*People* − 1).

Parabola Transformations—Book Sales

The publisher of the best-seller *The Secret Lives of Algebra Students* has asked 30 bookstores about the sales of that book. The stores reported how much they charged for each book and what their income was from their sales. The results were amazingly varied. Some stores charged very little for the book in order to sell more copies. Other stores charged a great deal for each book, making more on each sale but having fewer sales. The publisher turns to you for advice about what price a bookstore should charge to make the most income.

INVESTIGATE

1. The publisher gives you a scatter plot of the data. Open the Fathom document **Book Sales.ftm** to see this plot.
2. The publisher's representative believes that the graph fitting the data points would have an equation like $y = x^2$. Plot that function.

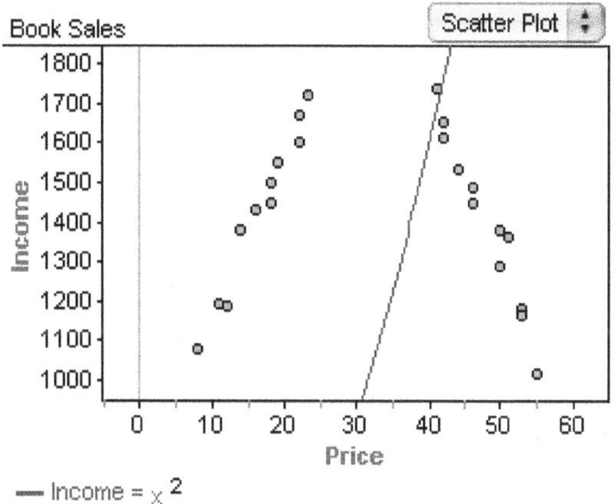

Q1 How would the graph of $y = x^2$ have to be transformed to fit the data points?

The shifts will be added or subtracted; the stretch will be a multiplier.

3. To build your transformed function without knowing what the exact transformations are, drag down sliders for *H_Shift*, *V_Shift*, and *V_Stretch*. Edit the formula $y = x^2$ to incorporate these three values.

Q2 What is your formula now?

4. Adjust the sliders until the transformed graph fits the data points well.

Q3 How does each slider affect the graph?

Q4 What function do you have for fitting the data?

5. A graph that fits the data points well will usually have the same number of data points above it as below. One good way to see the balance is to go to the **Graph** menu and choose **Make Residual Plot.** Each residual point shows the difference between a data point and the point predicted by the model. Those above the

horizontal line (positive) represent data points above the graph; the negative residuals represent data points below the graph. Adjust the sliders to have about the same number of positive and negative residuals.

In addition, the residual graph should not show a pattern. In the residual graph pictured here, you can see that the left side of the plot slants up and the right side slants down. This indicates that this model can be improved. Continue to adjust the model so the residuals don't show a pattern.

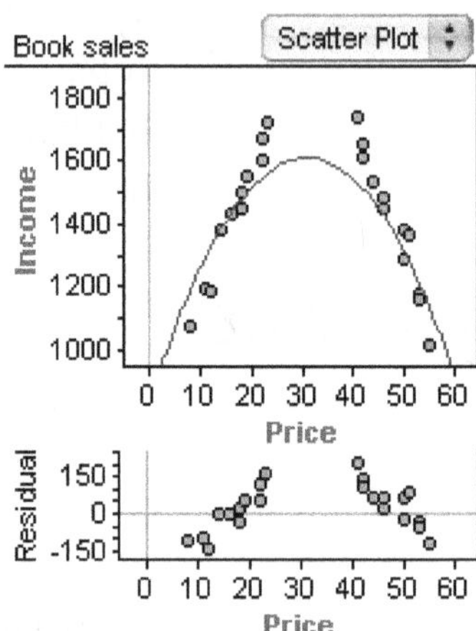

Q5 What's your function now?

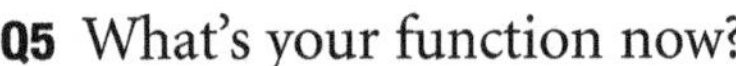

Q6 According to the function in Q5, what price should the publisher recommend so that bookstores can have the most income from the book? What is that income?

EXPLORE MORE

To fit linear data, you need only a vertical translation (shift) and a vertical dilation (stretch). However, it is sometimes easier to model if you include a horizontal translation. So the model $y = k + bx$ becomes $y = k + b(x - h)$. To fit parabolic data, you need both translations and one dilation. However, it is sometimes easier to model if you include both horizontal and vertical dilations. So $y = k + b(x - h)^2$ becomes $y = k + b\left(\frac{x - h}{a}\right)^2$.

1. Return to the data and alter your model to include a fourth slider, *H_Stretch.* Edit your formula to include all four sliders. Find at least four pairs of *V_Stretch* and *H_Stretch* that appear to fit the model well. Is there a relationship between these two values?

2. Select one of the points that is near or on your model. Subtract the vertex (h, k) from that point to calculate the change in x and the change in y. Use the change in x as *H_Stretch* and the negative of the change in y as *V_Stretch.* How does the model fit? Try this with some other points. Create a conjecture about finding values for *H_Stretch* and *V_Stretch.*

Parabola Transformations—Book Sales

Exploration

The publisher of the best-seller *The Secret Lives of Algebra Students* has asked 30 bookstores about the sales of that book. The stores reported how much they charged for each book and what their income was from their sales. The results were amazingly varied. Some stores charged very little for the book in order to sell more copies. Other stores charged a great deal for each book, making more on each sale but having fewer sales. The data are collected in the Fathom document **Book Sales.ftm.** The publisher asks you what price a bookstore should charge to make the most income.

The publisher's representative believes that the graph fitting the data points would have an equation like $y = x^2$. Use sliders to transform the graph of that function until it fits the data well enough to answer the publisher's question.

Parabola Transformations—Book Sales

Activity Notes

Objective: This activity uses quadratic data and Fathom sliders to deepen the understanding of transformations. Students will use sliders and residual plots to explore dilations of functions.

Student Audience: Algebra 1, Algebra 2

Activity Time: 20–30 minutes

Setting: Paired/Individual Activity or Exploration (use **Book Sales.ftm** for either setting)

Mathematics Prerequisite: Students know the basic transformation form of the quadratic, $y = k + b(x - h)^2$.

Fathom Prerequisites: Students can create and name sliders, set intervals, and plot a function and edit its formula.

Fathom Skills: Students will create and read a residual plot.

Notes: As you visit students, they may ask about the numbers, "If books sell for \$50, then why is this income not a multiple of \$50?" Explain that sales and special member discounts mean that not all books are sold for the list price. Help students connect the residual plot to the model by asking questions such as, Why is this residual point positive? or What if you increase the vertical shift? Q6 is important because it encourages students to connect the mathematical model to the problem situation.

Encourage students who complete the Explore More to symbolically pursue the question, Why do vertical stretches of a parabola appear to be horizontal shrinks?

For an Exploration: The Exploration form of the activity, for experienced investigators, does not mention residual plots. Student models should be near $Income = 1800 - 1.4(Price - 32)^2$.

INVESTIGATE

Q1 Answers will vary but should include at least "turning it over" (a vertical reflection) and a translation (shift) of the vertex to a point with coordinates approximately (35, 1700). Some students may mention stretching or shrinking as well.

Q2 $Income = V\_Shift + V\_Stretch(Price - H\_Shift)^2$, or an equivalent form. Be alert to student difficulties arising from using variables *x* and *y* rather than those shown in the graph, *Price* and *Income*.

Q3 Students may use *V_Shift* to raise and lower the graph, *H_Shift* to move left and right, and *V_Stretch* to flip the graph and change its proportions.

Q4 Answers will vary. One good function is $Income = 1800 - 1.4(Price - 32)^2$.

Q5 Same as Q4

Q6 The most income will occur at the parabola's vertex. For this function, that point is (32, 1800)—that is, the bookstores should sell the book for \$32 to receive an income of \$1800.

EXPLORE MORE

1. (*V_Stretch, H_Stretch*) or (a, b): $(\pm 1, -1.4)$, $(\pm 2, -5.6)$, $(\pm 3, -12.6)$, $(a, -1.4a^2)$. A vertical stretch is equal to -1.4 times the square of the horizontal stretch.

Expression for function

$$\text{Income} = V\_Shift + V\_Stretch\left(\frac{x - H\_Shift}{H\_Stretch}\right)^2$$

2. For the point (11, 1194.5), for example, $(32, 1800) - (11, 1194.5) = (21, 605.5)$. When *V_Stretch* is set to 21 and *H_Stretch* to -605.5, the curve will pass through the point (11, 1194.5). Conjecture: Once you have located a vertex, you can use any data point to determine the dilations needed to fit the curve through that point; subtract the vertex coordinates and use the change in *x* for the horizontal stretch factor and the negative of the change in *y* as the vertical stretch factor.

EXTENSION

If a bookstore finds that it sells 200 copies of a particular book at \$35.00 each and that every \$0.50 increase in the price of a book reduces the number of copies sold by 20, then what expression represents the bookstore's income as a function of price?

$$\begin{aligned} Income &= Price \cdot (\text{number sold}) \\ &= Price \cdot (200 - 20 \cdot (\text{number of \$0.50 increases})) \\ &= Price \cdot \left(200 - 20\,\frac{Price - 35}{0.5}\right) \\ &= 200\ Price - 40\ Price(Price - 35) \\ &= -40\ Price^2 + 1600 Price \end{aligned}$$

6

Investigating Higher-Degree Polynomials

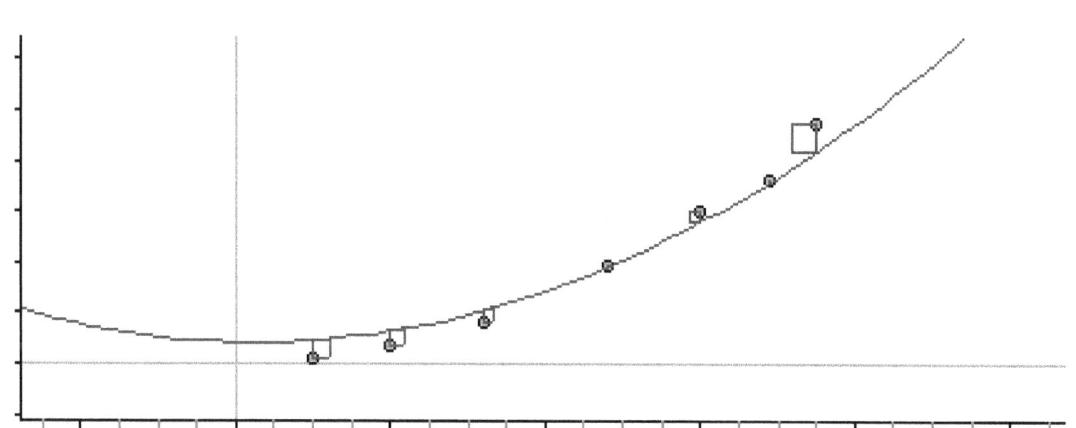

General Form Quadratic—Escape Ramp

On highways through mountainous regions, engineers often construct emergency escape ramps. Vehicles that have lost their brakes can use these ramps to come to a safe stop. To design one of these ramps, the engineers ran some tests with a truck traveling at various speeds. Your goal is to use their test data to figure out how long one of these ramps should be.

Q1 What factors do you think the engineers need to take into account?

INVESTIGATE

1. Open the Fathom document **Escape Ramp.ftm.** You will find a set of data that includes the speed of the truck, in miles per hour, at the start of the ramp and the distance to stop without using brakes, in feet.

For these data points, the engineers were using a ramp covered with sand and with a grade of 6% (that is, a slope of $\frac{6}{100}$). You find out that a ramp should be designed for a speed of 90 mph. The engineers didn't make a test at that high a speed, so you need to make a prediction.

2. Start by creating a scatter plot of the data, with *Speed* on the horizontal axis.

Because the data points appear to be curved, they might be fit by the graph of a quadratic function: $f(Speed) = a \cdot Speed^2 + b \cdot Speed + c$.

3. Drag down sliders for the coefficients *a*, *b*, and *c*. Plot the function with formula a • Speed² + b • Speed + c. To see the parabola, set the sliders near the position shown here.

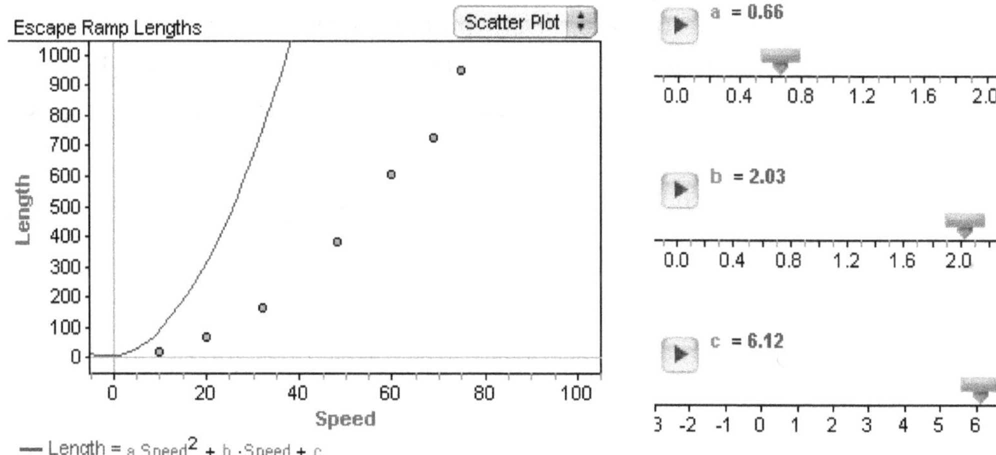

4. Because you need to make a prediction for a faster speed, enlarge your graph window to include speeds of at least 90 mph, as well as some negative values.

Q2 As you slide the value of c, how does the graph change? Include negative values of c. Explain your observations. Consider the location of the vertex and the shape of the parabola, as well as whatever else you see.

Q3 As you slide the value of b to positive and negative values, how does the graph change? Again, consider the vertex and the shape, among other things.

Q4 As you slide the value of a, how does the graph change?

Once you adjust the sliders so that the function graph fits the data points, you can determine the equation from the slider values.

Q5 What values of a, b, and c give a graph that fits the data points closely? You might choose **Graph | Show Squares** to help make a good fit.

Q6 How long should a ramp be to handle vehicles traveling at 90 mph?

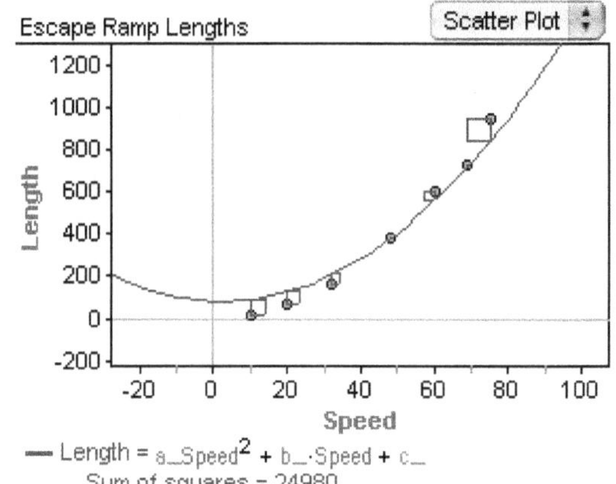

EXPLORE MORE

1. A quadratic function can also have the form $f(x) = a(x - h)^2 + k$. In this form, you can tell that the parent function $f(x) = x^2$ has been shifted h units horizontally and k units vertically and has been stretched or shrunk vertically by a factor of a. Use c to represent $f(x)$; algebraically write k in terms of a, c, and h. If a is fixed and you keep the value of $f(x)$ the same for some particular value of x (that is, you also fix x and c) while moving the slider for h, what is the path of the vertex?

2. Add units to the attributes in the table. You'll get error messages about incompatible units on the function or functions being graphed. To clear up these error messages, go to the sliders and enter appropriate units after the values.

   Fathom automatically changes ft/mph to s. Explain the relationship between ft/mph and seconds.

General Form Quadratic—Escape Ramp

Exploration

On highways through mountainous regions, engineers often construct emergency escape ramps. Vehicles that have lost their brakes can use these ramps to come to a safe stop. To design one of these ramps, the engineers ran some tests with a truck traveling at various speeds. You can find their data in **Escape Ramp.ftm.** Unfortunately, you need to plan a ramp for a speed they didn't test: 90 mph. Use sliders for the coefficients of a quadratic function to fit their data and decide the length of a ramp to handle a runaway speed of 90 mph.

General Form Quadratic—Escape Ramp

Activity Notes

Objective: Students will use Fathom and a real-world problem to dynamically explore how a graph is affected by the coefficients in the general form of a quadratic.

Student Audience: Algebra 1, Algebra 2

Activity Time: 25–40 minutes

Setting: Paired/Individual Activity, Exploration, or Whole-Class Presentation (use **Escape Ramp.ftm** for any setting)

Mathematics Prerequisites: Students understand points in a scatter plot, equation of a curve, and fitting a function to data in order to make a prediction.

Fathom Prerequisites: Students can make a scatter plot and adjust the graph window, plot and trace a function, create and name sliders, and show residual squares.

Notes: As you listen to students describe what they see in Q2–Q4, encourage complete descriptions by asking about the shape, even when it is not changing. Ask about the *y*-intercept if students don't mention it as they describe what they see as slider *a* is moved. Ask what happens when $a = 0$. If students see little change, encourage them to enlarge the scale on their sliders to see big changes. As students work on Q5, encourage them to set windows similar to those shown in step 3. Sliders in dynamic data software allow students to visualize the path followed by the parabola's vertex as coefficients change. The visualization can provoke the question "Why?"

For an Exploration: Students familiar with Fathom who do not need step-by-step guidance can use the Exploration. Their model and the approximation they make from their model will be close to answers for Q5 and Q6. As they refine their model, you might suggest the tool **Show Squares.**

For a Presentation: As you build the Fathom document from data, start with wide scales for the slider parameters so students can see clear differences as the sliders are moved. Draw out complete descriptions, letting several students contribute to those for Q2–Q4. For Q5, set the scales to measure in hundredths. As you **Show Squares**, ask students to describe what the squares are. Include Explore More 1 in the presentation. As you discuss Q6, you might ask, Is it reasonable to extend the linear data to fit trucks traveling at 90 mph? Students will have various opinions.

The engineers planning the ramp will need to know the maximum speed for which the linear relationship seen in the tests holds.

Q1 Answers will vary. Factors include the speed and weight of the runaway vehicle, the grade of the ramp, and the material from which the ramp is constructed.

INVESTIGATE

Q2 The parabola shifts up and down, as seen by the *y*-intercept (probably close to the vertex). The shape of the parabola does not change.

Q3 The vertex follows a path of a parabola opening down. The shape of the parabola remains the same.

Q4 The parabola closes and opens, corresponding to vertical stretches and shrinks (which are horizontal shrinks and stretches, respectively) as the absolute value of *a* increases and decreases. A negative value of *a* inverts the parabola, making it open downward. The *y*-intercept stays the same.

Q5 Answers will vary. Theoretically, the best quadratic fit is $Distance = 0.16Speed^2 - 0.09Speed + 2.45$. Student answers may vary more on values for *b* and *c*, which have less effect on the parabola through these data points. Ask whether it makes sense for the vertex to be at the origin. Of graphs with vertices at the origin, good fits have equations close to $Distance = 0.15Speed^2$.

Q6 Answers will vary. The models in Q5 give lengths of 1290 and 1215 ft, respectively.

EXPLORE MORE

1. $k = c - a(x - h)^2$. If *a* and *c* are fixed, the equation is a quadratic function of *h*, so the path is parabolic.
2. Slider *a* is in $\frac{\text{ft}}{\text{mph}^2}$, *b* is in $\frac{\text{ft}}{\text{mph}}$, and *c* is in ft.

$$1 \cdot \frac{\text{ft}}{\text{mph}} = \text{ft} \cdot \frac{\text{hr}}{\text{mi}} \cdot \frac{\text{mi}}{5280\text{ ft}} \cdot \frac{3600\text{ s}}{\text{hr}}$$

$$= 0.682\text{ s}$$

Units were not used in the main part of the activity to avoid confusion when Fathom automatically converts $\frac{\text{ft}}{\text{mph}}$ to seconds.

Factored Form Quadratic—Gravity

A football is kicked from a height of 2 ft and hits the ground 3.5 s later.

Q1 How high do you think the ball rises before beginning to fall?

Q2 What factors might determine that height?

Any object rising and falling due to gravity can be modeled with the quadratic function $y = ax^2 + bx + c$, where y is the height in feet and x is the time in seconds since the object was launched.

INVESTIGATE

You can set up a graph to show your view of the football field.

✓ Empty Plot
Function Plot

1. In a new Fathom document, drag from the shelf sliders for the coefficients a, b, and c. Drag a graph from the shelf. In the upper right corner of the graph, select **Function Plot** in the drop-down menu.

Make sure you click the multiplication sign between a and x^2 and between b and x.

2. To begin graphing the relationship between time and the height of the football, choose **Plot Function** from the **Graph** menu. Enter as an expression the quadratic function ax² + bx + c.

It might help to **Plot Value** 3.5 and **Plot Function** $y = 2$ as guides.

3. To make the parabola fit the data, adjust the values of a, b, and c until the graph represents a ball that is 2 ft high at time 0 s and 0 ft high at time 3.5 s. To make the parabola open downward, use negative values for a.

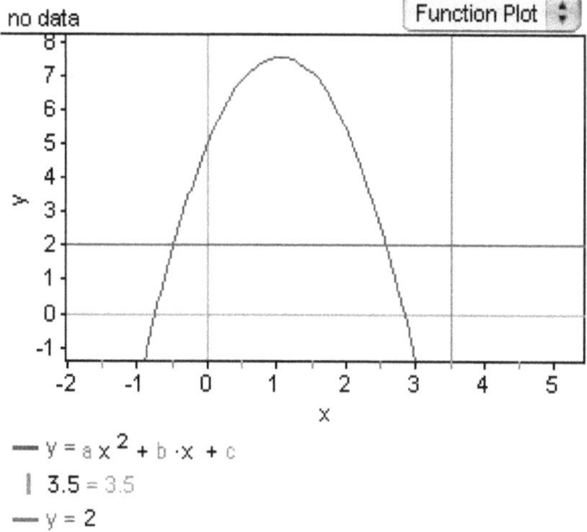

Q3 What values did you find for a, b, and c? What quadratic expression represents the height of the ball over time?

Q4 For your quadratic expression, how high does the ball rise?

Q5 On Earth, the value of a is -16. If you set a to -16, what values of the other sliders make the height 2 ft at 0 s and 0 ft at 3.5 s?

Q6 How high does the ball rise with these settings?

4. The *factored form* of the quadratic function is $y = a(x - z_1)(x - z_2)$. Drag sliders from the shelf for z_1 and z_2—for names, you might enter **z1** and **z2**. Replace the equation you've graphed with one in this factored form.

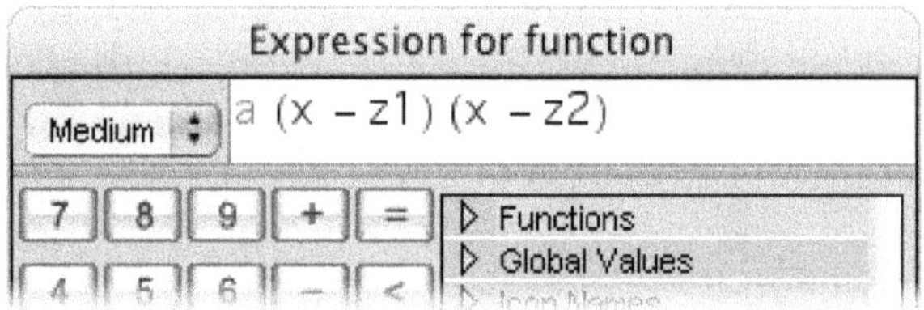

Q7 Still using $a = -16$, what values for z_1 and z_2 give a graph that passes through (0, 2) and (3.5, 0)?

Q8 How do the values of z_1 and z_2 relate to the graph?

Q9 Why are z_1 and z_2 called the *zeros* of the function?

EXPLORE MORE

Look at how the sum of the two zeros relates to *a* and *b* and how the product of the zeros relates to *a* and *c*.

1. Go back to graphing $ax^2 + bx + c$. Experiment with changing the value of *a*. Keep track of the values of *a*, *b*, and *c* and the zeros of each function. How do the zeros relate to the values of *a*, *b*, and *c*?
2. Research how the values of *a* and *b* relate to kicking a football. If a ball kicked on Earth from a height of 2 ft hits the ground in 3.5 s, what was its initial upward velocity and how high did it go?

Factored Form Quadratic—Gravity

Objective: Students will see relationships among the coefficients of a quadratic equation in general form, the factored form of the equation, and the function's zeros (the x-intercepts of its graph). Students are given the factored form and, by using sliders, discover how this form relates to the graph.

Student Audience: Algebra 1, Algebra 2

Activity Time: 25–35 minutes

Setting: Paired/Individual Activity

Mathematics Prerequisites: Students can read a function graph and locate x-intercepts.

Fathom Prerequisites: Students can use sliders, plot a function referring to a slider, and trace a function graph.

Fathom Skills: Students learn how to make a function plot.

Notes: As you talk with students, don't let them confuse the path of the football, also a parabola, with the function they graph in step 2—a function that relates the height of the football to time, not height and horizontal distance traveled. A wide variety of answers are possible on Q3 and Q4. You might ask students to check their own answers to make sure that with $c = 2$, $b = -3.5a - 0.57$. If students ask why, suggest that the quadratic formula would help them verify the relationship between a and b. As students are working on Q7, suggest that they enlarge the graph and ensure that the scale for z_1 is set to measure in thousandths. During a class discussion after the activity, or as students share their answers, make sure they can explain the answers to Q8 and Q9. A function's zeros are the x-intercepts of its graph, or the solutions to its equation.

Q1 Answers will vary considerably. You need not comment on them.

Q2 The primary influences are the force of gravity and the initial upward velocity of the ball when kicked. The angle of the kick affects the upward speed. Air resistance is another factor, which is ignored in this model.

INVESTIGATE

Q3 The many correct answers each have $c = 2$ and a negative. Theoretically, because one of the zeros of the function is at $x = 3.5$ for $c = 2$, $3.5 = \frac{-b \pm \sqrt{b^2 - 8a}}{2a}$. Therefore, b will be approximately equal to $-3.5a - 0.57$. For example, $-x^2 + 2.93x + 2$.

Q4 Answers depend on the equation from Q3. The maximum height is $c - \frac{b^2}{4a}$. For the example from Q3, the y-coordinate of the vertex is 4.15.

Q5 $b \approx 55.4$, $c = 2$

Q6 About 50 ft

Q7 The z-values are about -0.036 and 3.5.

Q8 The z-values are the x-intercepts of the graph.

Q9 The zeros of a function are the values at which the function is 0.

EXPLORE MORE

1. Students can keep track of their data in a Fathom table, letting Fathom calculate the sum and product. The sum of the zeros is $-\frac{b}{a}$; their product is $\frac{c}{a}$.
2. The initial upward velocity of the ball is b; c is the ball's initial height; a is acceleration toward Earth due to gravity. The ball kicked on Earth would have initial upward velocity of 55.4 $\frac{\text{ft}}{\text{s}}$ and would reach a height of about 50 ft.

Vertex Form Quadratic—Protecting Wildflowers

The school ecology club has permission to fence in a region along a riverbank to protect some endangered wildflowers that grow there. The club has enough money to buy 220 feet of fencing. It decides to enclose a rectangular space. The fence will form three sides of the rectangle, and the riverbank will form the fourth side.

Rather than trample down wildflowers, the club makes some rectangles along the side of the school to determine which has the largest area.

Q1 How large an area do you think can be fenced on three sides using 220 ft of fence?

INVESTIGATE

1. Open the document **Wildflowers.ftm** to see the data from the club's experimentation. Create a scatter plot of (*Width*, *Area*).
2. The data appear parabolic. Because you want to find a vertex of the parabola, it would be useful to use the vertex form: $Area = a(Width - h)^2 + k$. Create sliders for a, h, and k, and enter the function.

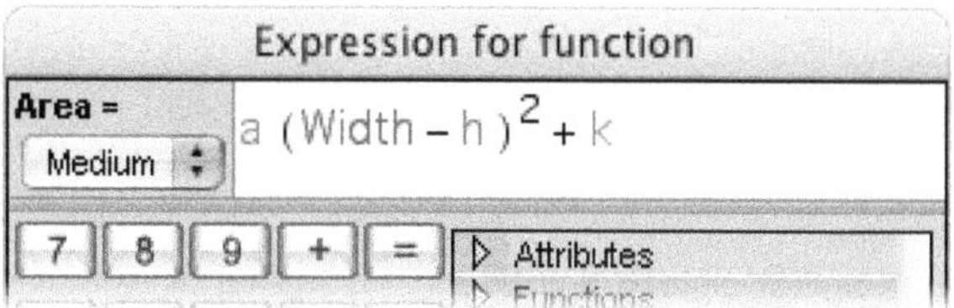

You get an error message that the units are incompatible.

Area = a (Width – h)2 + k (#Units incompatible#)

3. To eliminate that message, you'll need to put units on each slider. Decide which sliders use feet and square feet.

Q2 What are the units for each slider?

4. Experiment with the sliders until you have a good fit for the data.

Q3 What are the values of the sliders?

Q4 How do these values relate to the problem?

Q5 How do these values relate to the graph?

Q6 Why is this form of the equation called the *vertex form*?

Q7 What are the dimensions of the largest rectangle?

EXPLORE MORE

1. What if the rectangle is not along the riverbank but is enclosed only by fence? What length and width will give the maximum area?
2. One club member suggests that they would get more area if they used a trapezoid with 45° base angles instead of a rectangle. Use the data from the Trapezoid Area collection to discover whether this is correct.
3. Could forming the fence into a different nonrectangular shape enclose more area?

Vertex Form Quadratic—Protecting Wildflowers Exploration

The school ecology club has permission to fence in a region along a riverbank to protect some endangered wildflowers that grow there. The club has enough money to buy 220 ft of fencing. It decides to enclose a rectangular space. The fence will form three sides of the rectangle, and the riverbank will form the fourth side.

Rather than trample down wildflowers, the club makes some rectangles along the side of the school to determine which has the largest area. To see the data from the club's experimentation, open the document **Wildflowers.ftm.**

To model the data, it might be easy to start with the vertex of the parabola; you can use a function in vertex form: $Area = a(Width - h)^2 + k$, where the vertex is at (h, k). Use sliders for a, h, and k to find a function that fits the data and determine the dimensions of the largest rectangle.

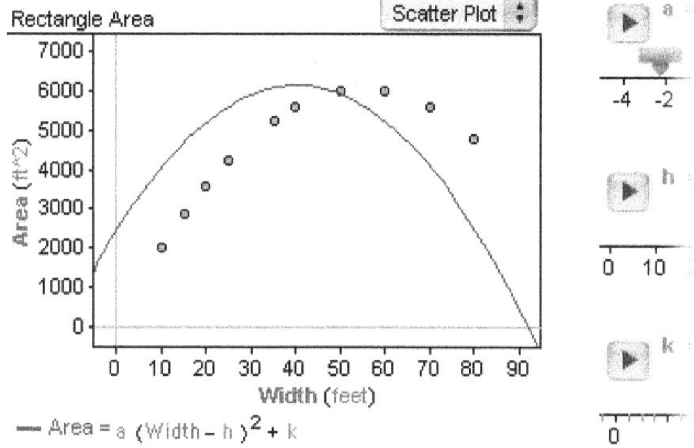

EXPLORE MORE

1. What if the rectangle is not along the riverbank but is enclosed only by fence? What length and width will give the maximum area?
2. One club member suggests that they would get more area if they used a trapezoid with 45° base angles instead of a rectangle. Use the data from the Trapezoid Area collection to discover whether this is correct.
3. Could forming the fence into a different nonrectangular shape enclose more area?

Vertex Form Quadratic—Protecting Wildflowers Activity Notes

Objective: Students will create a strategy for modeling quadratic data based on the vertex-dilation equation. Fathom's sliders and units are used to discover the roles of the values used in the vertex form of the quadratic.

Student Audience: Algebra 1, Algebra 2

Activity Time: 30–45 minutes

Setting: Paired/Individual Activity, Exploration, or Whole-Class Presentation (use **Wildflowers.ftm** for any setting)

Mathematics Prerequisites: Students understand the graph of a function.

Fathom Prerequisites: Students can use sliders, create a scatter plot, and plot a function based on slider values.

Notes: Help students who are struggling to find the units for the sliders by asking about units when numbers are added or subtracted. Suggest they write the equation and then check that the units are the same on both sides of the equal sign. The open-ended version of the Exploration allows students to build on Fathom skills developed earlier. Some students might benefit from showing a residual plot as well. As students report values for h and k in Q3, check that they include units. If you can give students more than 25 minutes for this activity, many will be able to complete the Explore More questions. You might ask students who complete those additional questions to share their results with the class.

For a Presentation: Encourage discussion of Q2. You'll also want to present the Explore More questions.

Q1 Answers will vary.

INVESTIGATE

Q2 The a slider needs no units, the h slider should be in feet, and the k slider should be in square feet.

Q3 $a = -2$, $h = 55$ ft, $k = 6050$ ft^2

Q4 Values h and k give the desired width and maximum area. Value a does not directly relate to the problem.

Q5 Values h and k give the position of the parabola's vertex. Value a relates to the proportions of the parabola.

Q6 It is called the *vertex form* because the coordinates of the vertex are part of the equation.

Q7 The theoretical maximum area occurs when the width is one-quarter of the total 220 ft and the length is twice that: $55 \text{ ft} \cdot 110 \text{ ft} = 6050 \text{ ft}^2$.

EXPLORE MORE

1. The maximum area of a rectangle with a fixed perimeter occurs when the rectangle is a square. Students can use this fact to solve the original problem as well: Imagine reflecting the rectangle across the riverbank to have a rectangle with twice as much fencing (440 ft). Form that into a square, so each side has one-quarter of 440 ft. On the dry side of the riverbank, each width is half that long, or one-eighth of 440 ft, and the length is one-quarter of 440 ft.
2. A trapezoid will give greater area. The maximum area has the approximate dimensions of width 60 ft and bases of 50 ft and 170 ft, for an area of 6600 ft^2.
3. Regular polygons with the same perimeter will have more area if they have more edges. Of all figures with the same perimeter, the circle has the most area. Against a riverbank, the semicircle has more area than any other figure with the same perimeter.

EXTENSION

The first sentence of Explore More 3, known as the *Isoperimetric Theorem,* was known by the ancient Greeks. Suggest that students research the history of this theorem and its proof.

The Quadratic Formula

The quadratic formula says that you can calculate the zeros of the quadratic function $f(x) = ax^2 + bx + c$ by finding the two values of $\frac{-b \pm \sqrt{b^2 - 4ac}}{2a}$.

CREATE A MODEL

In a new Fathom document, drag down sliders for the coefficients *a*, *b*, and *c* and make a function plot of the formula $ax^2 + bx + c$. Also drag down sliders for zeros z_1 and z_2 and give the sliders formulas for the zeros, as described by the quadratic formula.

INVESTIGATE

Explore what happens to the graph and the zeros as the values of *a*, *b*, and *c* change. What can you say about the graph when z_1 and z_2 have values and when they have domain errors? Explain.

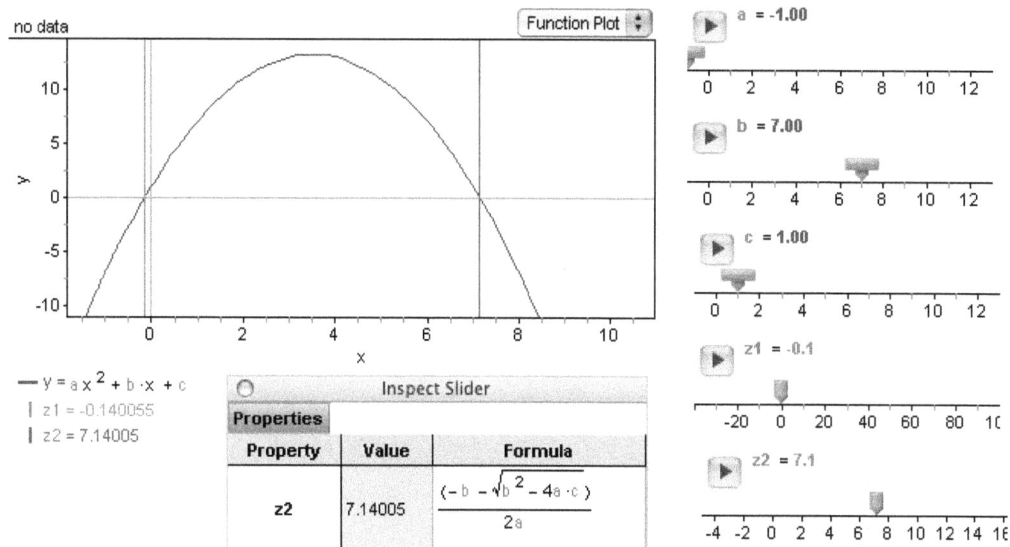

Objective: Students will explore the roles of the coefficients of a quadratic function in the locations of the zeros, as given by the quadratic formula, and in the location of the x-intercepts of the function's graph.

Student Audience: Algebra 1, Algebra 2

Activity Time: 25–35 minutes

Setting: Paired/Individual Exploration

Mathematics Prerequisites: Students understand the graph of a quadratic, associating zeros of a function with the x-intercepts of the function's graph.

Fathom Prerequisites: Students can use sliders, make a function plot with a formula based on sliders, and assign formulas to sliders.

Notes: This open-ended exploration allows students to see how the values of the zeros of a quadratic function, as given by the quadratic formula, relate to the location of the parabolic graph. In particular, you want them to see that the zeros become undefined, with Fathom indicating #Domain error#, when the graph has no x-intercepts. Patterns seen in Fathom can motivate the question, How do a, b, and c relate to each other when there are zeros and when there aren't? The question can be answered algebraically by considering the discriminant, $b^2 - 4ac$.

Parabola—Solar Oven

You want to build a parabolic solar oven to be fit into a frame. Your design is to have a dish 50 cm deep and 150 cm across at the level of the frame. You need a formula for the dish to give you the measurements you need to construct the oven.

INVESTIGATE

On the new graph, change Empty Plot to Function Plot. Then choose **Graph | Plot Value** and **Graph | Plot Function** to draw lines at $x = 150$ and $y = -50$.

1. Open a new Fathom document and drag down a graph for a function plot. Use the dish measurements to help set the window. You might plot values to show the dimensions of the dish.

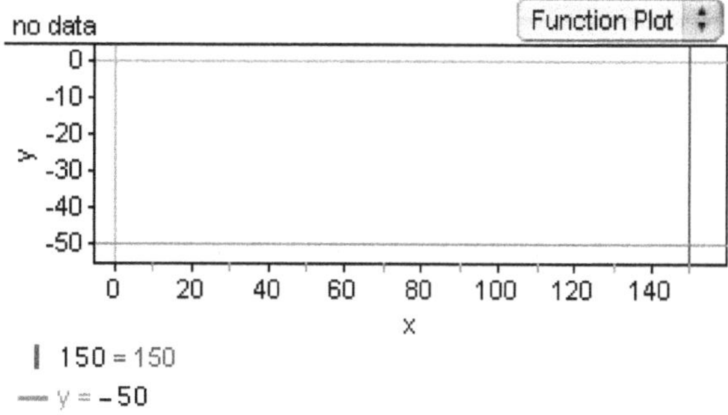

2. Because the dish design is parabolic, its formula will be $y = ax^2 + bx + c$. Drag down sliders for the coefficients a, b, and c.

Adjust the slider scales so that a, b, and c can take on values close to zero.

3. Plot the function with formula $ax^2 + bx + c$ and adjust the sliders to place the curve in the box. (Don't worry about a perfect fit just yet.)

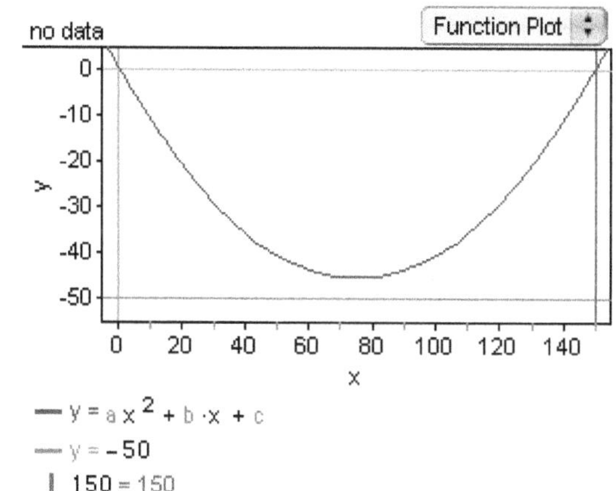

The quadratic formula, $\frac{-b \pm \sqrt{b^2 - 4ac}}{2a}$, will give you the location of the x-intercepts of any quadratic curve of the form $y = ax^2 + bx + c$.

Q1 What are the approximate x-intercepts of your current graph, and what are the x-intercepts of the curve you want?

Another way to write the quadratic formula is as two fractions: $\frac{-b}{2a} \pm \frac{\sqrt{b^2 - 4ac}}{2a}$. This form will help you fit your graph in the box.

Double-click on the slider's thumb to open its inspector and enter the formula.

4. Create a new slider, enter the name V1, and give it a formula of $\frac{-b}{2a}$. Then, with the graph selected, choose **Graph | Plot Value** to draw a vertical line on your graph at *V1*.

Q2 Where is this line on your parabola? (Check your hypothesis by moving the sliders and watching the line.)

Q3 Where will this line need to be on the finished parabola?

5. Set the expression $\frac{-b}{2a}$ equal to your answer to Q3 and solve for *b*. Enter this formula for slider *b*.

Q4 As you slide the value of *a*, how does the graph change?

Q5 What values of *a*, *b*, and *c* give a graph that best fits the box? You might need to adjust the scales to look at small intervals.

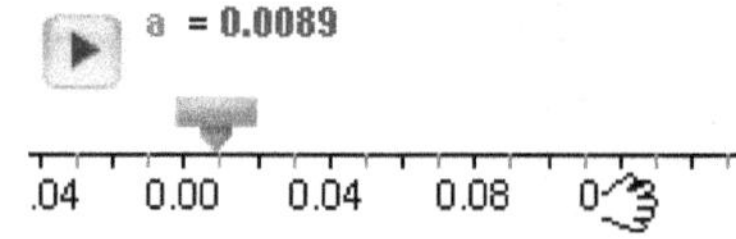

If you get the error message #Units incompatible#, you may have forgotten to click on the multiplication sign between 4, *a*, and *c* or between 2 and *a*.

6. Create two more sliders, using the values of *a*, *b*, *c* and the formulas $\frac{-b+\sqrt{b^2-4ac}}{2a}$ and $\frac{-b-\sqrt{b^2-4ac}}{2a}$.

Q6 Aided by the values of the new sliders, can you improve your model?

Q7 To help construct the parabola, how high off the bottom of the oven are the points that are 40 cm from the center horizontally?

EXPLORE MORE

Experiment with the *a*, *b*, and *c* sliders until the values of the quadratic formula sliders give an error message. What is different about this graph, as compared with graphs that don't give error messages?

Parabola—Solar Oven

Objective: Students will use a quadratic equation to model a solar oven and will discover some of the relationships between the quadratic formula and the graph.

Student Audience: Algebra 1, Algebra 2

Activity Time: 25–35 minutes

Setting: Paired/Individual Activity or Whole-Class Presentation

Mathematics Prerequisites: Students understand the graph of a quadratic and associate zeros of a function with the x-intercepts of the function's graph.

Fathom Prerequisites: Students can create sliders, make a function plot with a formula based on sliders, and assign formulas to sliders.

Notes: The margin note for step 3 will help students who are having trouble getting the curve into the box. As you listen to students working, look for an understanding that $\frac{-b}{2a}$ gives the horizontal component of the vertex and that the symmetry of this curve means the two intercepts must be equidistant from this value. As students work on Q6 ask, How will the values of the new sliders help? The distance $\frac{\sqrt{b^2 - 4ac}}{2a}$ is both added and subtracted from the center to find the intercepts. For students familiar with function notation, you might ask them to use function notation as they write their answer: $f(35) = f(115) \approx 35.8$.

For a Presentation: Set the values of a and c near 0 and the value of b near -1 to get the curve to roughly fit the box. Ask several students to describe what they see at Q4. As the sliders are created in step 6, ask what these slider values will be when the curve fits the box. After tracing the curve to find the answer to Q7, you might extend the question to other numbers, such as 30 cm from the center or 20 cm from the top.

INVESTIGATE

1. The following answers assume that the dish will be below the x-axis and to the right of the y-axis. The screen captures in the activity suggests this placement. However, there are other valid ways to begin the activity.

Q1 The intercepts of student curves will vary. The desired intercepts are 0 and 150 cm.

Q2 It passes through the vertex of the parabola.

Q3 The x-coordinate of the vertex should be at 75 cm.

5. $75 = \frac{-b}{2a}$, $b = -75 \cdot 2a = -150a$

Q4 The x-value of the vertex is always at 75 cm. The y-value of the vertex increases and decreases.

Q5 Best answer is $a = \frac{2}{225}$, or about 0.00889; $b = -\frac{4}{3}$, or about -1.333; and $c = 0$.

Q6 Same as Q5

Q7 About 35.77 cm below the top, or about 14.22 cm above the bottom

EXPLORE MORE

A parabola with errors instead of values is a parabola that does not cross the horizontal axis.

Binomial Products—Sales and Profits

You have developed a great-tasting nutrition drink. You sell it in 12-packs to 20 retail markets in your area. Some of the discount stores resell the 12-packs at a low price so as to sell a large number of packs. Some health clubs sell drinks individually at a high price and sell only a few packs. You have decided to sell your own product at a local festival, but you need to choose a price.

Q1 Is it better to sell many drinks at a low price or a few at a high price? Explain your ideas.

Test your opinions by collecting data on last month's sales at each outlet.

INVESTIGATE

1. Open the Fathom document **Sales.ftm**. You will find a case table of the selling price per pack from each outlet, the profit they made on each pack, and the total sales for the previous month. You can start your research by looking for any patterns in these values. Create scatter plots for *Sell_Price* versus *Profit_per_Pack* and *Sell_Price* versus *Packs_Sold.*

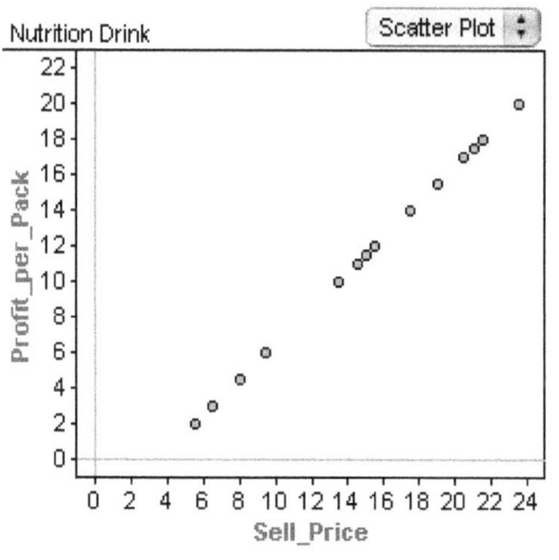

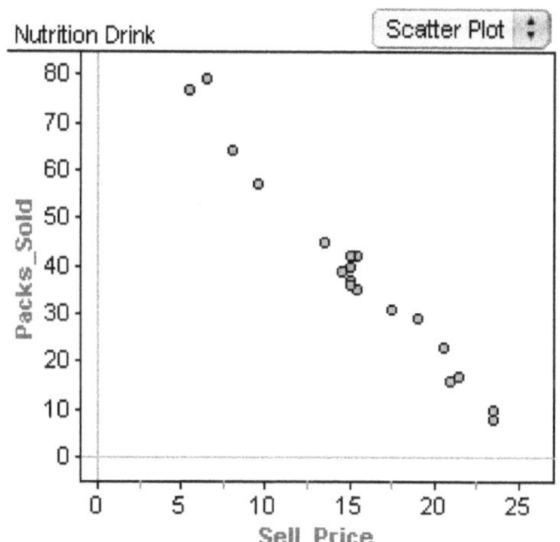

2. Find the best line of fit modeling each of these graphs.

Q2 What model did you use for the first graph? What can you learn from its intercepts with the axes?

Q3 Give the model for the second graph and explain what the slope in this model tells you.

Q4 How would you calculate how much money Albert's Market made from this product last month?

Nutrition Drink

	Outlet	Sell_Price	Profit_per_Pack	Packs_Sold
1	Albert's Market	13.5	10	45

3. What you want to know is which stores made the most money. Add a new attribute in the case table with a name like *Total_Profit* and enter a formula to calculate this value.

Q5 What formula did you use? Which outlet made the greatest profit?

4. Because the best price according to the model may not be one of the prices any outlet charged, you will want to look for a formula. Create a third graph to study how profit relates to the selling price. Use sliders to help you find a model to fit these data.

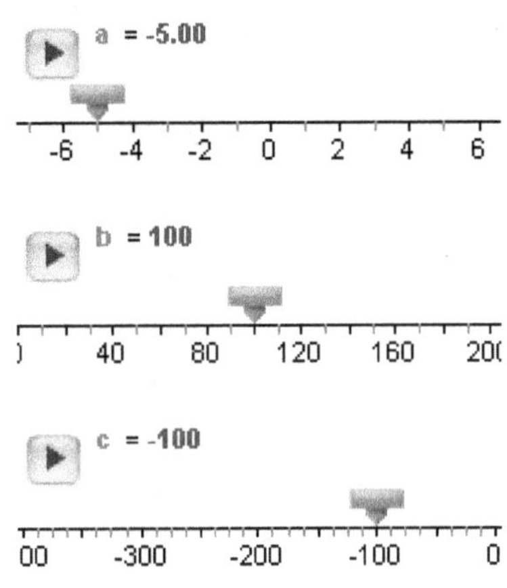

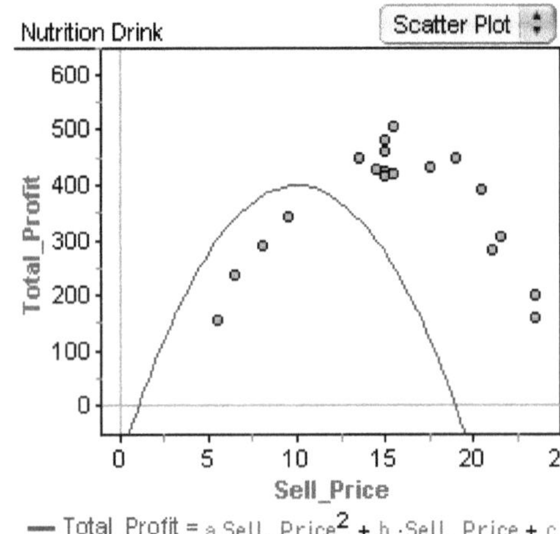

Q6 What model did you find to fit these data? According to that model, what price should you charge at the festival, and what profit will you receive?

Now that you have solved the problem one way, you wonder whether using algebra can give you a solution without using sliders. You decide to compare the three graphs.

Q7 What are the horizontal intercepts of the three models? Explain any patterns you see.

Q8 Multiply the right sides of your answers to Q2 and Q3. Explain any patterns you see.

Q9 How could you have found the model for *Sell_Price* as a function of *Total_Profit* without sliders? What solution would you have gotten?

EXPLORE MORE

1. You have decided to go into the vegetable-growing business. You plant carrots. When these carrots are young, they are very sweet, but they get less tasty as they continue to grow in size. Each day you leave them in the field, they increase in size. Create a case table for the collection Carrots (scroll down in the Fathom document **Sales.ftm**) and use what you have learned to find the best day to harvest your crop for the highest profit. The values given are based on previous years and are only good approximations of this year's crop. A function for *Profit* in terms of *Time* will help you find the best theoretical values.
2. Explore the Walleye data (also in **Sales.ftm**) to determine for the fishery the best age at which to sell fish to make the highest profit.

Binomial Products—Sales and Profits

Objective: Students will explore how linear models can give information, both graphical and symbolic, about the quadratic model that is their product.

Student Audience: Algebra 1, Algebra 2

Activity Time: 30–40 minutes

Setting: Paired/Individual Activity or Whole-Class Presentation (use **Sales.ftm** for either setting)

Mathematics Prerequisites: Students can multiply binomials.

Fathom Prerequisites: Students can create a scatter plot, use movable lines, add attributes and edit formulas, use sliders, add function plots to a graph, and trace functions.

Notes: Step 2, Q2, Q3, and Q6 give students a chance to find a line (or curve) of fit and interpret the meaning of each function's terms for this problem situation. Students are gaining experience applying the process of finding a mathematical model to fit a situation, solving the model, and then interpreting the result back into the problem situation. Q7 and Q8 give students further experience with looking for patterns—doing mathematics.

For a Presentation: Ask several students to interpret the meaning of the constants and the coefficients in the lines of fit. Before you create the graph in step 4, ask students what shape they think the points on the scatter plot will have.

Q1 Answers will vary widely. You need not reach consensus at this time.

INVESTIGATE

2. Students may use movable lines, one of the built-in regressions, or the equation through two representative points. If they write the equation in point-slope form, encourage them to change it to intercept form to facilitate later calculations.

Q2 $Profit\_per\_Pack = -3.5 + Sell\_Price$, exact values may differ slightly. The vertical intercept is the per-pack wholesale cost to the retailer. Each item (pack) costs each store \$3.50. The horizontal intercept gives sales that would yield a profit of 0. Selling packs at \$3.50 would return no profit.

Q3 $Packs\_sold = 96 - 3.72 Sell\_Price$, exact values may differ slightly. The slope is the rate at which the number of sales decreases as the price increases. The retailer gets 3.72 fewer sales for each dollar increase in price.

Q4 Multiply \$10 per item times 45 items sold to get \$450 profit.

Q5 $Total\_Profit = Profit\_per\_Pack \cdot Packs\_sold$. Don's Beverage #1 and #3 made a profit of \$504 for the month.

Q6 $Total\_Profit = -3.72 Sell\_Price^2 + 109 Sell\_Price - 336$, exact values may differ slightly. The best price is about \$14.65 per pack (or \$1.22 each), with the expectation of selling about 40 packs for a profit near \$462. To avoid dealing with pennies, \$1.25 each is a good price for the festival.

Q7 Graph 1: intercept at $x = 3.5$ (representing zero profit). Graph 2: intercept at 25.8 (representing zero sales). Graph 3: intercepts at 3.5 and 25.8 (representing zero profit for either of these reasons). The zeros of the profit function are the zeros of its factors.

Q8 The product should be something like that in Q6. The profit function is the product of the other two functions.

Q9 The product of the two linear expressions is a quadratic whose zeros are those of the linear functions. The problem could have been solved by graphing the product of the two linear functions, to get a price of \$14.65 with a profit near \$462.

EXPLORE MORE

1. A good model to fit (*Time, Weight*) is $Weight = -180 + 5.23 Time$. A good model to fit (*Time, Price*) is $Price = 3.4 - 0.0298 Time$. The profit model is the product of these two: $Profit = -0.156 Time^2 + 23.15 Time - 612$. The best time to sell is at 74 days, making a profit of \$246.85.
2. A good model for (*Age, Weight*) is $Weight = -0.75 + 0.212 Age$. A good model for (*Age, UnitPrice*) is $UnitPrice = 17 - 0.329 Age$. The profit model is the product of these two: $Profit = -0.070 Age^2 + 3.85 Age - 12.8$. The best age to sell is at 27.5 days, yielding a profit of \$40.14.

Binomial Expansion—Flipping Coins

After you flip a coin, either the heads side or the tails side shows.

Q1 If you flip three coins, do you think it's more likely that you will see two heads or three heads?

Fathom can help you simulate the flipping of coins, so you can experiment with many flips without taking much time or wearing out your fingers.

INVESTIGATE

One way to simulate flipping coins is to set up a collection with two members, representing the outcomes of heads and tails. To flip three coins, take a random sample collection three times.

1. Open the Fathom document **Coin Flip.ftm.** You will see a Coin collection with an enlarged window, allowing you to see its two members: one Head and one Tail. The repeated draws from the Coin collection are in a different collection, Flips. At this point, the Flips collection contains the results of three coin flips. Click on **Sample More Cases** in the Flips collection to see the results of flipping three coins again.

Q2 Keep track of the results as you simulate flipping the three coins eight times. How many of those times do two coins show heads? How many of those times do three coins show heads?

Q3 Answer Q2 for another eight simulations.

The document contains a third collection, which will keep track of the results for you. The results are called *measures,* and this collection is named Measures from Flips.

To speed up the simulation, make sure animation is not on.

2. Open the inspector for Measures from Flips and use it to collect 1000 measures.

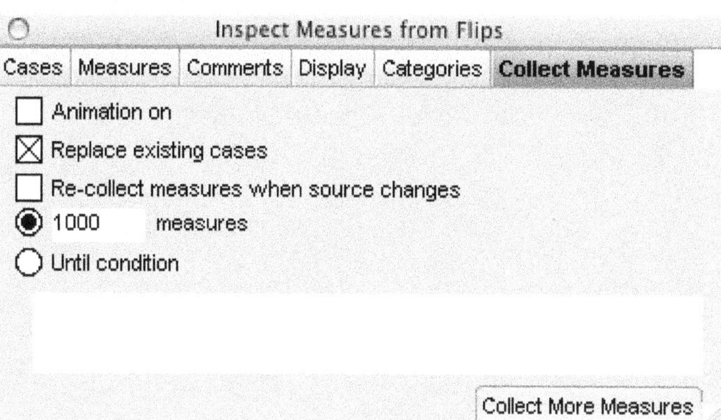

This graph was created by dragging the *Result* attribute from the **Cases** panel of the Measures from Flips inspector to the horizontal axis of a new graph.

3. Create graphs and summary tables to use this information to help you answer the questions.

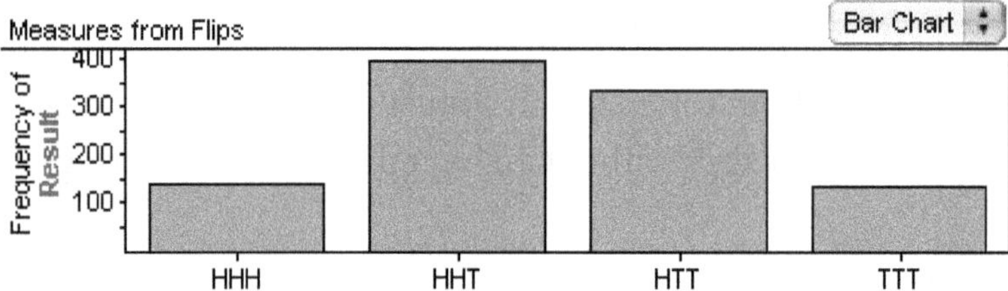

Measures from Flips

Result	HHH	112
	HHT	387
	HTT	386
	TTT	115
Column Summary		1000

S1 = count ( )

To create this summary chart from a different sample of 1000 flips of three coins, highlight the measures collection and drag a summary table from the object shelf. Open the measures collection inspector, click the **Cases** tab, and drag and drop *Result* on the down arrow in the summary table.

Q4 How likely is three heads? Two heads and one tail? One head and two tails? Three tails?

Q5 To find the connection between coin flipping and algebra, multiply $(H + T)(H + T)$, and then multiply the result by $(H + T)$ again.

Q6 Combine the like terms after you have multiplied. What is $(H + T)^2$? How do the coefficients relate to the coin flips?

You have seen the result for three flips. Does it apply to other numbers of flips?

4. To try for four flips, open the inspector for the Flips collection and change the number of cases to **4**. Simulate flipping the four coins 16 times, recording the number of heads each time.

Make sure you open the inspector with the title Inspect Flips.

Inspect Flips

Cases | Measures | Comments | Display | Categories | **Sample**

☐ Animation on ☒ With replacement
☒ Replace existing cases
☐ Collect new sample when source changes
◉ 4 cases
○ Until condition

Sample More Cases

When the Flips collection is updated to simulate flipping four coins, Measures from Flips will measure the number of heads (and tails) in four coins.

Q7 What were your results? Repeat another 16 times. What did you get this time? Combine your results with those of classmates and find the average numbers for 16 flips.

Q8 Multiply $(H + T)$ by the expansion you found in Q5. Combine the like terms after you have multiplied. How do the coefficients relate to the coin flips?

Q9 Calculate $(H + T)^5$ and guess at what fraction of the time flipping five coins will produce four heads.

5. To test your conjecture, adjust your simulation as you did in step 4 so that you are sampling five flips. Simulate 32 flips several times.

Q10 Use algebra to conjecture what fraction of the time flipping six coins will produce two heads. To test this pattern, simulate 64 flips several times.

EXPLORE MORE

Scroll down in the **Coin Flip.ftm** document to find a graph labeled Triangle_1. Drag the corner to enlarge the graph. The rows of the triangle have 1, 2, 4, 8, 16, and 32 points, respectively. All but the last row are arranged as the coefficients of a binomial expansion. Look for patterns and then arrange the 32 points in the last row to fit the pattern.

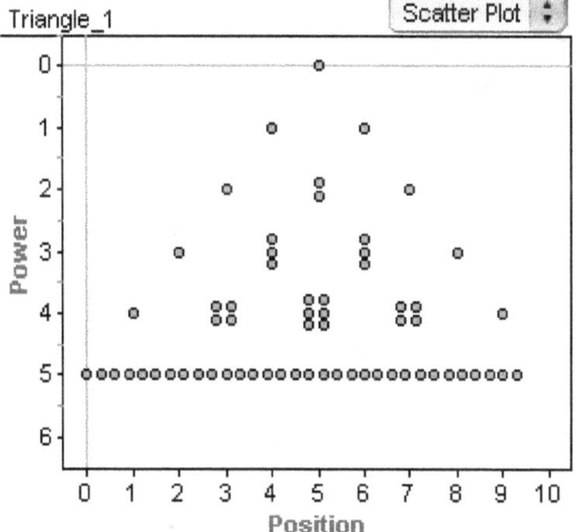

Extend the pattern further by writing numbers instead of arranging dots. Do you recognize this famous number triangle?

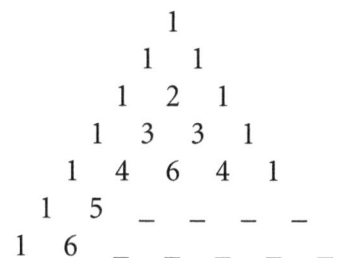

1
1 1
1 2 1
1 3 3 1
1 4 6 4 1
1 5 _ _ _ _
1 6 _ _ _ _ _

Binomials Expansion—Flipping Coins

Objective: Students will explore how patterns in binomial expansions relate to flipping coins.

Student Audience: Algebra 1, Algebra 2

Activity Time: 15–25 minutes

Setting: Paired/Individual Activity (use **Coin Flip.ftm**)

Mathematics Prerequisites: Students can multiply polynomials.

Fathom Prerequisites: Students have worked with collections.

Notes: As you introduce the activity or talk with students, help them understand two powerful tools of Fathom—the ability to collect samples and the ability to collect measures from those samples. Collecting samples involves cases selected at random from the original collection and assembled into a new collection. In this activity, the original collection is the two possible outcomes for a coin flip, and the sample collection is one flip of three coins. To see the trend in the long run, you take many samples. The more samples you take, the closer the result will be to the result in the long run.

You need a tool to keep track of the count of heads for each flip of three coins. A measure is a calculation based on the sample collection. For example, you can count the number of heads in each sample and save that information—in this case, it is saved in Measures from Flips. When Fathom collects measures, it re-collects the sample, calculates the formula created for each measure (counts the number of heads), and then builds a new collection with the results. The attributes in this new collection are the measures—here, count of heads—and each case represents a measure taken from a different random sample—here, one flip of three coins.

Ask students who finish early to take measures from 1000 simulations of flipping four coins, five coins, or more. Students who are interested might examine and explain the formulas for the measures, which are taken as the sample is collected and then sent to the measures collection.

During sharing, the class can look at and combine several results for Q4 by making a table in Fathom for class results and then using Fathom to combine those results to get another picture of the probability in the long run. Taking samples with the animation on can help students see what is happening—where the numbers are coming from. You might show sampling from coins five times with animation on. When collecting a large number of samples, however, turn off the animation.

Q1 Answers may vary substantially. Consensus will come through the activity.

INVESTIGATE

Q2 Answers will vary. The number of times two coins show heads will be about three times the number of times all three coins show heads.

Q3 Same as Q2

Q4 There will be approximately three times as many two heads and a tail or two tails and a head as there are three coins the same.

Q5 $HHH + 3HHT + 3HTT + TTT$, or $H^3 + 3H^2T + 3HT^2 + T^3$

Q6 The coefficients are roughly the same as the counts. Point out that the term $3H^2T$ means that two H's and one T are multiplied in three different ways when expanding the binomial; two heads and one tail can occur HHT, HTH, or THH.

Q7 For a large number of trials, the number of times various numbers of heads will appear are approximately 1, 4, 6, 4, 1. With only 16 trials, the actual results may vary so greatly from the theoretical result that the patterns may not appear. Combining data for the class will help reveal the pattern.

Q8 $HHHH + 4HHHT + 6HHTT + 4HTTT + TTTT$, or $H^4 + 4H^3T + 6H^2T^2 + 4HT^3 + T^4$

Q9 $H^5 + 5H^4T + 10H^3T^2 + 10H^2T^3 + 5HT^4 + T^5$. Four heads will appear about 5 times out of 32.

Q10 $H^6 + 6H^5T + 15H^4T^2 + 20H^3T^3 + 15H^2T^4 + 6HT^5 + T^6$. Two heads will appear about 15 times out of 64.

EXPLORE MORE

The bottom row of the dots should be arranged in groups of 1, 5, 10, 10, 5, 1. The pattern is Pascal's triangle; each number is the sum of the two numbers above it. The last row pictured is 1, 6, 15, 20, 15, 6, 1.

Common Factor—Acceleration

Acceleration due to gravity affects how fast objects fall, how projectiles travel, and how much force is needed to launch rockets. But acceleration differs around the world. Your goal in this activity is to find the value of that acceleration in a particular place.

Because falling objects travel so fast that measurements are difficult, the famous scientist Galileo devised a way to find the acceleration due to gravity by using objects rolling on a ramp, or an incline. Here you'll be using some data collected on an incline.

INVESTIGATE

1. Open the Fathom document **Acceleration.ftm**. You will see a table of distance and time data from an experiment.

To collect these data, an incline was constructed by raising by 11 cm one end of a 2.44 m table. A probe for measuring distance and time was placed at the bottom. A can was placed on the incline above the probe and given a shove upward. The measuring device recorded the distance between the can and the probe at various times.

2. Create a scatter plot of *Distance_from_Probe* versus *Time.*

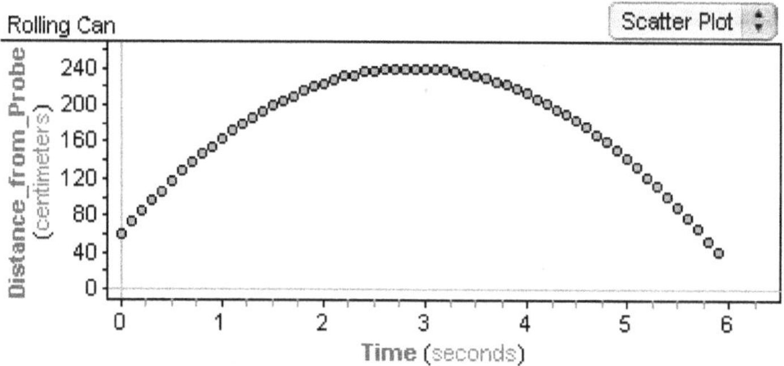

Q1 What type of model would you use for these data? How far from the probe was the can when the data collection started?

3. You could find a model for these data with sliders, but here you'll use an approach based on factoring. If z is a zero of the function (represented on the graph by a horizontal intercept), then $(Time - z)$ is a factor of the function. To make this factor easy to use, adjust the graph so that 0 is a zero of the function. Create a new variable that measures the distance from the start of the data collection rather than the distance from the probe. Name this new attribute

something like *Distance_from_Start*, enter the formula shown here, and change the dependent variable on the graph.

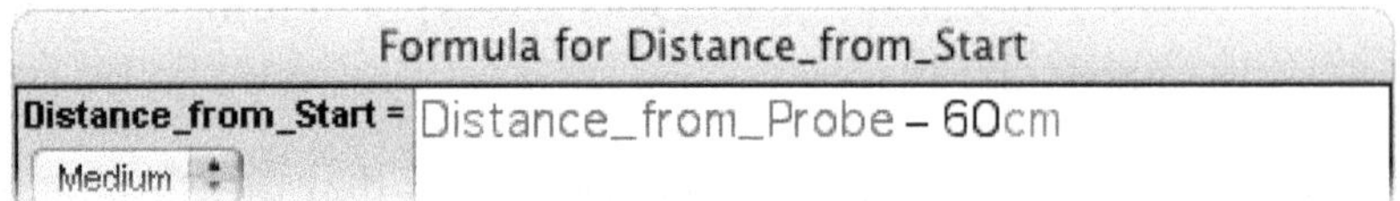

Q2 How does this graph differ from the first graph? Explain.

Now, ($Time - 0$), or simply *Time*, is a factor of the *Distance_from_Start* function. If you divide *Distance_from_Start* by *Time*, you should get linear data, which are easy to model. $\frac{Distance\_from\_Start}{Time}$ even has a meaning: It is average speed.

4. To get the linear data, create a new attribute called *Average_Speed*, give it the formula shown here, and make a scatter plot of *Average_Speed* versus *Time*.

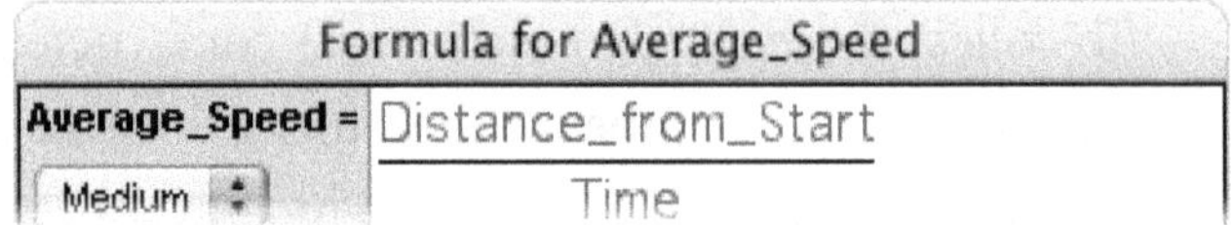

Formula for Average_Speed
Average_Speed = Distance_from_Start / Time
Medium

Q3 Describe this graph.

Q4 What equation fits the data in this graph?

5. To see how this model can be used to find a quadratic model, start with the equation $d = at^2 + bt + c$ and subtract c from both sides. Then divide both sides by t.

Q5 What is the resulting equation, and how does this new equation relate to your graph in Q2 and its equation?

Q6 Using your answer to Q4 and the factor *Time*, find a quadratic equation to model the original data. Check your equation by graphing it on a scatter plot of *Distance_from_Probe* versus *Time*.

Q7 The value of a in $d = at^2 + bt + c$ is the acceleration of the can along the incline. Because the incline rises 11 cm over 244 cm of length, the acceleration a is related to the acceleration due to gravity, which we'll call g, by the formula $a = \frac{1}{2}g\left(\frac{11}{244}\right)$. Solve this equation to find the acceleration due to gravity where the experiment was performed.

Q8 Explain how factoring helped you to find a model for the quadratic.

EXPLORE MORE

Many snowplows have large cone-shaped bins filled with sand and salt to spread on snow-covered roads. It is quite time-consuming to refill the bins, but it is also time-consuming to have to stop in the middle of a route to return because you have run out. You wish to build a model that the road crews could use to estimate how far they can travel before they are out of sand/salt. In the collection called Salt Truck, the road crews have provided you with 20 data points collected where they have recorded the height of the sand and distance the plow went before it was empty. Describe the steps taken to model these data with a function in the same way you proceeded in this activity. What is the function?

Common Factor—Acceleration

Objective: Students will use factoring to find a quadratic model of some real-world data involving motion on an incline.

Student Audience: Algebra 1, Algebra 2

Activity Time: 15–25 minutes

Setting: Paired/Individual Activity (use **Acceleration.ftm**)

Mathematics Prerequisites: Students can solve equations.

Fathom Prerequisites: Students can create a scatter plot, add attributes and edit their formulas, use movable lines to find a line of fit, and add function plots.

Notes: As you talk with students, encourage them to keep the units in mind; distances in centimeters, speeds in centimeters per second, acceleration in centimeters per second squared. The ratio of distances is unitless; it results from dividing centimeters by centimeters. As students work on Q3, they might note that the points near the y-axis, where both the time and the distance are small, show the greatest error. They can ignore those points when they fit the graph. The text before step 4 is essential for students to understand; they know one factor of the *Distance_from_Start* quadratic function; by dividing by *Time*, they will find the other factor. Ask several students to share their answer to Q6 and the explanation they wrote for Q8.

INVESTIGATE

Q1 These data points seem parabolic, so a quadratic model should be best. The first data value (at time = 0) is 60 cm. The can was 60 cm from the probe when the data collection started.

Q2 It has the same shape, but it is 60 units lower and now passes through the origin.

Q3 This graph shows a linear relationship.

Q4 Answers may vary, but they'll be close to the least squares line of
Average_Speed = -22.1 cm/s^2 *Time* + 126 cm/s

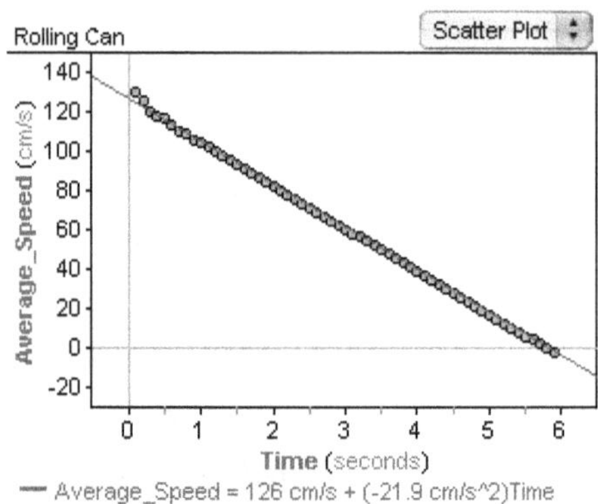

Q5 $d = at^2 + bt + c$

$d - c = at^2 + bt$

$d - c = t(at + b)$

$\frac{d-c}{t} = at + b$

The ratio $\frac{d-c}{t}$ gives average speed, so the line on the graph of *Average_Speed* versus *Time* has a slope of a and a vertical intercept of b. The equation for this line will be one factor in the quadratic function that fits the *Distance_from_Start* function.

Q6 $-22.1t^2 + 126t + 60$, or
$-22.1\frac{\text{cm}}{\text{s}^2}\textit{Time}^2 + 126\frac{\text{cm}}{\text{s}}\textit{Time} + 60$ cm

Q7 $a = \frac{1}{2}g\left(\frac{11}{244}\right)$

$-22.1 = \frac{1}{2}g\left(\frac{11}{244}\right)$

$-22.1 = \frac{11}{488}g$

$-980\frac{\text{cm}}{\text{s}^2} \approx g$

Q8 Many observations are possible here. A remark that the two expressions $at^2 + bt + c$ and $t(at + b) + c$ are the same would show good understanding.

EXPLORE MORE

The height is 0 at about 11.5 mi, so set up a new variable at which *Height* will be 0. Divide the new variable, *Reduced_Height*, by *Height* to get *Average* and find a line of fit for the data (*Height*, *Average*).

$$\textit{Reduced\_Miles} = \textit{Miles} - 11.5$$

$$\textit{Average} = \frac{\textit{Reduced\_Miles}}{\textit{Height}}$$

The least-squares line through the data for (*Height, Average*) is *Average* = 4.90*Height* − 10.5. So one approximate model is *Miles* = 4.90*Height*^{2} − 10.3*Height* + 11.5.

Polynomial Factoring—Maximum Area

Mathematical analysts in business and industry collect data and create models to find maximums, such as the maximum yield or the maximum profit, and minimums, such as the minimum waste or the minimum cost. Simple problems like folding a sheet of paper to find the largest triangle are used to train analysts for their work.

On a sheet of 8.5-by-11-inch paper, mark each inch from top to bottom along the left 11-inch edge of the sheet. Fold the upper right corner to one of the marks and crease the paper. There is now a right triangle of a single thickness in the upper left corner of the page, above the part of the edge that is folded. The two legs of the triangle are along the side and the top of the paper.

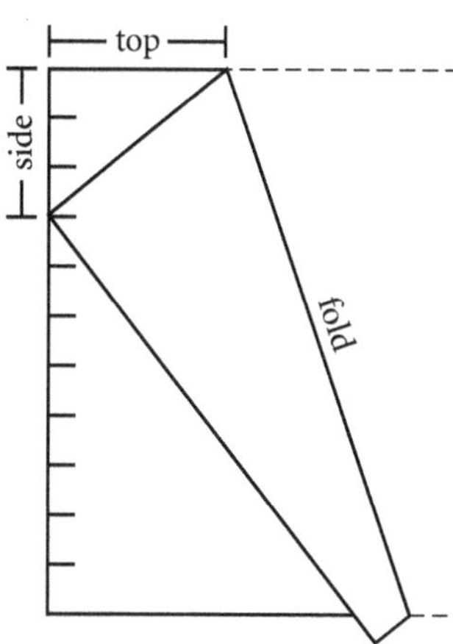

Q1 Which mark do you believe will result in the triangle with the largest area?

INVESTIGATE

The analyst must find the exact position of the fold to give the largest area in that triangle.

1. Open a new Fathom document. Drag down a case table and create the attributes *Side* and *Top*. Record in the table the lengths of the triangle's leg as you move the top right corner to marks along the left edge.

Q2 How many marks can you actually use? Explain.

2. Create a new attribute for *Area*, using the formula $0.5 \cdot Side \cdot Top$. Your goal is to find the exact position for the fold that makes the triangle the largest. To help you see the data, create a scatter plot of *Area* versus *Side*.

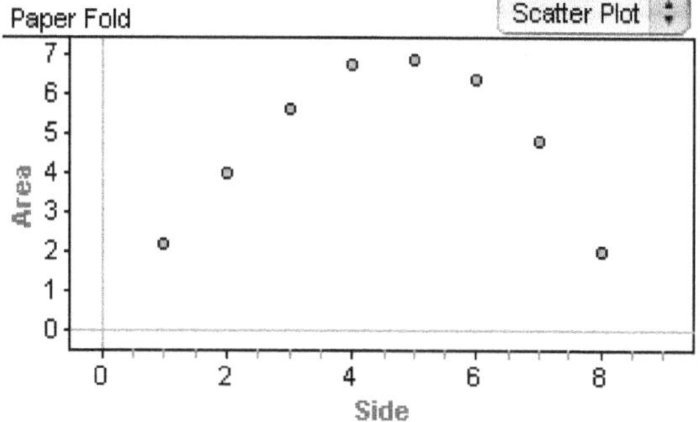

The graph looks somewhat quadratic. The graph of a quadratic function has symmetry, with the highest point halfway between the horizontal intercepts.

Q3 Think about how you gathered these data. Where should the horizontal intercepts be? That is, which values of *Side* would give you 0 area? What point is halfway between the two side lengths with no area?

Q4 Do you believe these data are actually quadratic? Why or why not?

The easiest type of model to find is linear. Often in statistics, you look for ways to change the data in order to "unbend" the curves, then you reverse the process to bend the line after you have found a model. This sequence is called *linearization.*

If z is a zero of a function, meaning the horizontal intercept of the graph, then $(x - z)$ is a factor of the function. Because you know two intercepts of this graph, you know two factors. You can create a data set of lower degree, and therefore one that is more linear, by dividing the data by one factor.

3. Create a new attribute called something like *Area_factored* and give it the formula of *Area* divided by one of the factors you know.

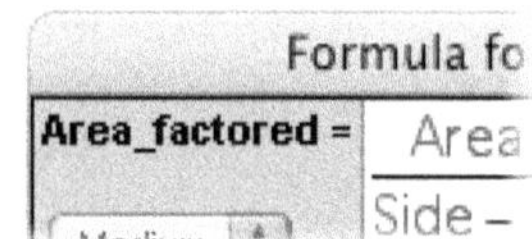

4. The original data will be quadratic if the values of *Area_factored* are linear. Make a scatter plot of *Area_factored* versus *Side.*

Q5 Is this graph linear or curved? Is it increasing, decreasing, or both?

5. Because the data are not yet linear, divide *Area_factored* by the other factor, creating *Area_factored_twice.* Create a scatter plot of this new attribute versus *Side.*

Q6 Use a movable line to find a linear model for the data points in this graph.

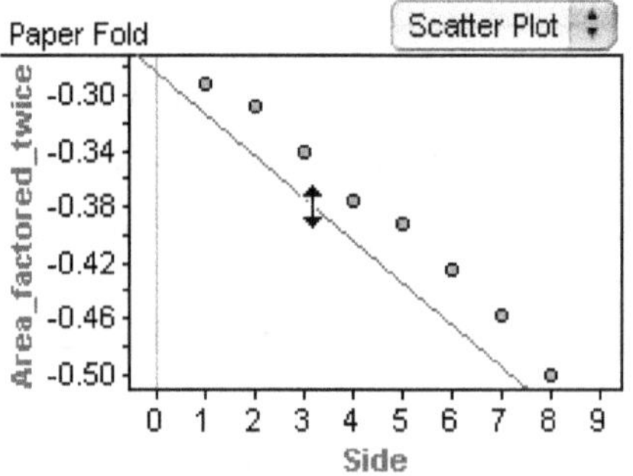

6. To find the model you're seeking for area, work backward. Start with the equation you found in Q6 and multiply it by each of the factors you used to

make the data linear. Test this model by plotting it as a function on the scatter plot of *Area* versus *Side*.

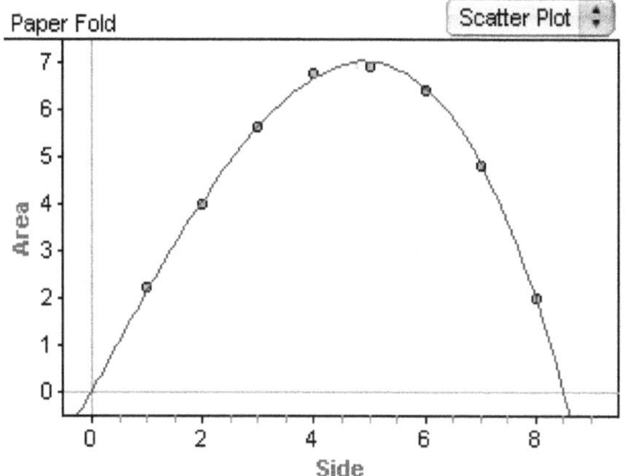

Q7 What is your model for the area?

Q8 What does tracing the graph tell you about how to fold the paper to get a triangle of maximum area? According to your model, what is that area?

EXPLORE MORE

You found models for *Area_factored_twice* and for *Area*. How can you adjust the model for *Area_factored_twice* to get a model for *Area_factored*?

Polynomial Factoring—Maximum Area

Objective: Students will use factoring as a part of a process of modeling a third-degree polynomial. Students will explore the relationships among intercepts, zeros, and factors as they maximize area in a paper-folding activity.

Student Audience: Algebra 1, Algebra 2

Activity Time: 20–35 minutes

Setting: Paired/Individual Activity or Whole-Class Presentation

Optional Document: Area.ftm

Mathematics Prerequisites: Students can solve equations.

Fathom Prerequisites: Students can create a data table with numeric and formula attributes, edit attribute formulas, use movable lines, and add function plots.

Notes: This activity can start with the collection of data using a sheet of 8.5-by-11–inch paper and a ruler, or you can save time and use the premeasured data in **Area.ftm.** You might start by demonstrating how to fold the paper and showing the location of the triangle that students need to measure. If your time is limited and you start with the data in **Area.ftm,** first demonstrate what is being measured. If students are confused, go back to the physical model, perhaps labeling the side and the top. As you visit working pairs, find one group who divided first by (*Side* − 0) and another that used (*Side* − 8.5). Ask both pairs to be prepared to share.

For a Presentation: If you only have access to one computer with presentation capability, you can still ask students to gather the data. Start a table in Fathom to enter each group's *Side* and *Top* measurements for each inch mark, then use the class average for the presentation. You might plot the value $x = 8.5$ and talk about Q4 and Q5.

Before the student running the computer shows the graphs in steps 4–6, ask what students expect to see. Ask the Explore More question.

Q1 Students will likely pick the 4 in. mark or 4.25 (halfway between 0 and 8.5). This is a good guess, but it is not so exact as the one they will derive later.

INVESTIGATE

Q2 At inch marks below 8 in., there is no triangle.

Q3 At 0 in. and at 8.5 in. If students have trouble, encourage them to think about a triangle with no area.

Q4 If the graph were symmetric, then the maximum would be at 4.25 in., but it is not. The data are probably not quadratic.

Q5 The graph is not linear. It will be decreasing if students divide by the factor (*Side* − 0), but it will be increasing if they divide by the factor (*Side* − 8.5).

Q6 Answers will vary, depending on the accuracy of students' original measurements. The model is something like $y = -0.03Side - 0.25$.

Q7 Using the above answer to Q6, $Area = (Side - 0)(Side - 8.5)(-0.03Side - 0.25)$.

Q8 Answers will vary. The maximum area occurs when *Side* is about 4.9 in., giving an area of more than 7 in.^{2}

EXPLORE MORE

Area_factored = *Area_factored_twice* · *Side* or *Area_factored* = *Area_factored_twice* · (*Side* − 8.5), depending on the last factor divided out

7

Simulating Probability

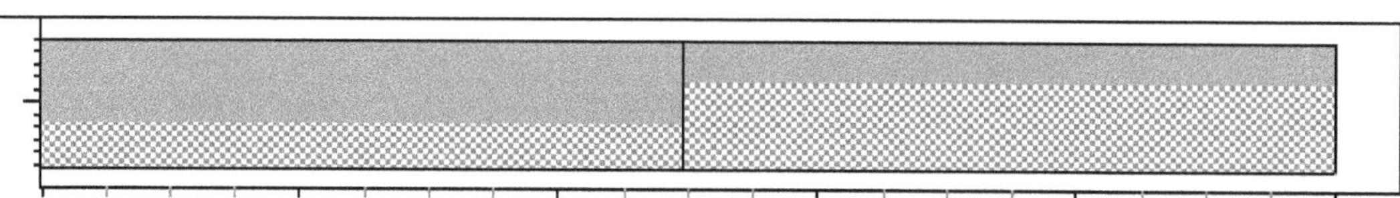

Probability—Dice Games

Two friends, who like to play games of dice, decide to play a new one. On each turn, Player 1 rolls two dice 24 times, while Player 2 waits. If one of the 24 rolls comes up with double 1s (called a *double ace*), then Player 1 wins. If Player 1 doesn't win, then Player 2 takes a turn by simply rolling four dice. If at least one 1 (an *ace*) is rolled, then Player 2 wins. If Player 2 doesn't win, then Player 1 takes a turn again. And so on.

Q1 Which player do you guess will win the game most often?

INVESTIGATE

Fathom can simulate chance, such as the outcome of rolling a die. Once you have set up the simulation for one of the turns, you can quickly simulate many of those turns.

Open the collection's inspector to enter attributes and their formulas. Choose **Collection | New Cases** to create 24 cases.

1. Open a new Fathom document. For Player 1's turns, drag a collection from the shelf and name it something like Two Dice. Create two attributes, *Die_1* and *Die_2*. Because the dice are rolled 24 times, make 24 cases.
2. To simulate the chance results of rolling a die, give each attribute the formula randomInteger(1,6). This formula randomly produces an integer from 1 to 6, simulating the result of a die roll.
3. Create a scatter plot with *Die_1* on one axis and *Die_2* on the other.

Q2 Can you tell from the scatter plot whether the turn wins the game? Explain.

Note the keyboard shortcut to simulate a turn even faster: Ctrl+Y (Win) or ⌘+Y (Mac).

4. To simulate another turn, you can have Fathom select new random numbers to simulate new dice rolls. Simply choose **Rerandomize** from the **Collection** menu.

Q3 Simulate ten turns and record the number of turns that win the game. Did more turns win than not?

Fathom can make many turns for you and record the results. To set this up, you'll need to create a variable that indicates whether or not the turn wins.

5. Add a new attribute called something like *Double_Ace*. For this attribute to tell whether the turn won the game, give it the formula (Die_1 = 1) and (Die_2 = 1). Drag a new graph into the document and drag *Double_Ace* to the horizontal axis.

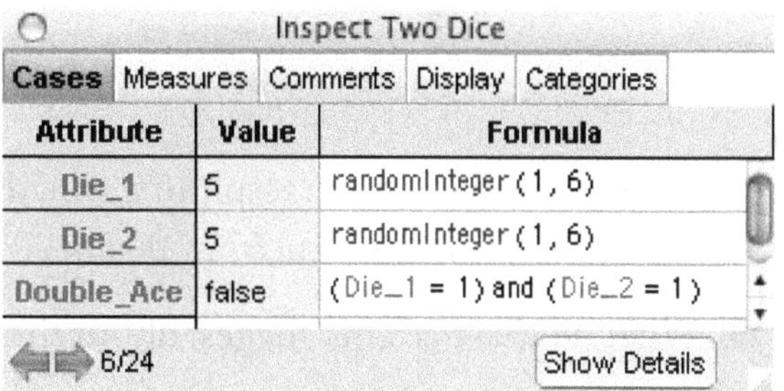

Inspect Two Dice
Cases
Measures
Comments
Display
Categories
Attribute
Value
Formula
Die_1
5
randomInteger(1, 6)
Die_2
5
randomInteger(1, 6)
Double_Ace
false
(Die_1 = 1) and (Die_2 = 1)
6/24
Show Details

Q4 What type of graph does Fathom create? How can you tell from the graph whether the turn won the game? Use **Rerandomize** to simulate another ten turns. What percentage of these turns won the game?

The next step in getting Fathom to make many turns for you is to create a count of the number of times a turn contains a double ace. The count is called a *measure* in Fathom.

6. In the collection inspector, click on the **Measures** tab. Create a new measure, perhaps named *Game_Won*, and enter a formula to count each time there is at least one time in 24 rolls when *Double_Ace* is true.

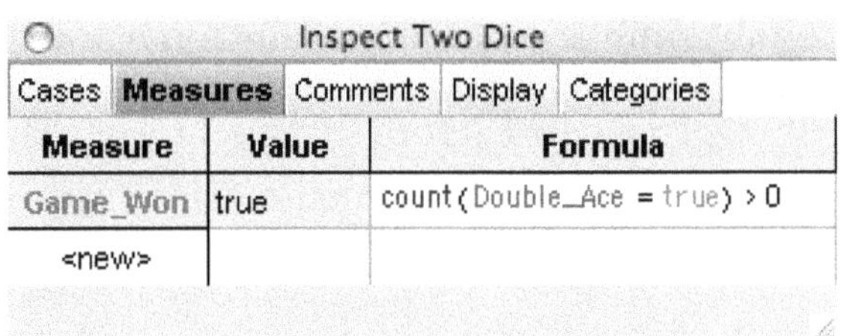

Simulate a turn of Player 1 by clicking **Rerandomize** on the Two Dice collection. In the inspector's **Measures** panel, you should see that *Game_Won* says "true" every time the bar chart indicates a true and the scatter plot shows a (1, 1) among the 24 dots.

7. Test the new measure by replaying the game ten times while keeping track of the number of times the outcome is true.

Q5 How many times did *Game_Won* show true? What percent of the time did the turn win the game?

8. To collect the counts on 100 turns, select the Two Dice collection. Choose **Collection | Collect Measures.** A new collection, named Measures from Two Dice, is created.

9. Open the inspector of the Measures from Two Dice collection and set the properties as shown here. Then click on **Collect More Measures** so that 100 turns are simulated.

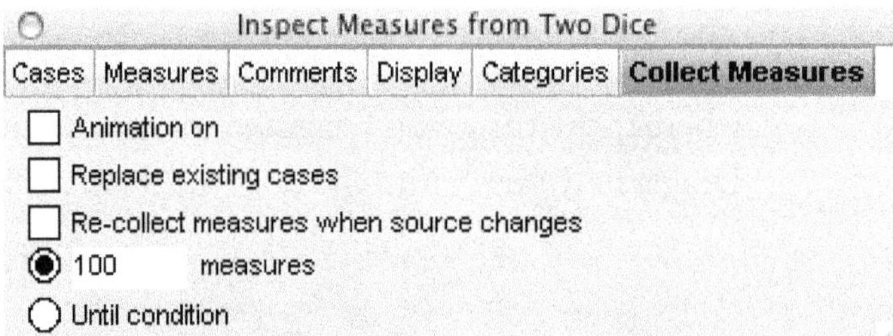

To drag the attribute from the Measures from Two Dice inspector, first click on the **Cases** tab.

10. To see how many turns result in a win, drag a summary table from the shelf. Drag the *Game_Won* attribute from this measures collection to the summary table.

Q6 What do these counts suggest about the probability that a Player 1 turn will win a game?

Q7 What might you do to improve your approximation of the probability? Do that, and give a new estimate of the probability of winning this game. Are you surprised?

The Four Dice collection will have only one case, not 24.

11. Using steps 1–9 as a starting point, create a process to model Player 2's turns, in which four dice are rolled.

Q8 What's the approximate probability that a Player 2 turn will win the game?

EXPLORE MORE

1. Suppose the rules were changed so that Player 1 wins if doubles of any type (not just 1's) come up. How many rolls of two dice should the player be given to make the game as fair as possible? How do you know?
2. A third friend joins the game and wants to roll three dice, winning if three aces are thrown. How many times should this player roll three dice so that the probability of winning is as close as possible to 0.5?

Objective: Students will use Fathom to simulate a game based on a historical question that led to probability as a branch of mathematics.

Student Audience: Algebra 1, Algebra 2

Activity Time: 50–60 minutes

Setting: Paired/Individual Activity or Whole-Class Presentation (use **Dice Games Present.ftm**)

Optional Document: Dice Games.ftm

Mathematics Prerequisites: Students have had some experience with probability.

Fathom Prerequisites: Students can create a new collection, add attributes and enter their formulas, create a scatter plot, and use inspectors.

Fathom Skills: Students learn to create measures and enter their formulas, collect measures, and use a summary table.

Background: You might introduce the activity with this history: A gambling question started a rich correspondence between two mathematicians, Pascal and Fermat, that founded probability theory as an accepted branch of mathematics. The seventeenth-century gambler and nobleman who started their conversation was Chevalier de Méré. De Méré was a better gambler than a mathematician, but he reasoned that the two games—getting four chances to roll an ace, which has a probability of $\frac{1}{6}$, and getting 24 chances to roll a double ace, which has a probability of $\frac{1}{36}$—would have the same likelihood of winning, because 4:6 = 24:36. However, he made money playing the one game and lost money playing the other, so he asked Pascal and Fermat about it.

Notes: If students are used to working with Fathom, they will probably not get bogged down in the many steps needed to create the simulation. However, if some students are struggling, you might suggest they start with **Dice Games.ftm,** which has the collections and their measures already built. Students starting with the Fathom document can skip steps 1, 2, and 7 and part of steps 5, 6, 8, and 10. On step 11, they can scroll down to see the collections to simulate Player 2's turn. As you visit groups, make sure students understand the steps and the formulas. You might ask a student to explain the formula in step 6, or ask a student to describe the source of the numbers in a summary table. Ask a group who has built a simulation for Player 2 to share their results with the class. You could extend the lesson to teach the ideas of probability needed to make the theoretical calculations. The probability of *not* getting a double ace on a roll of two dice is $\frac{35}{36}$, so the probability that a Player 1 turn does not win is $\left(\frac{35}{36}\right)^{24}$. Hence, the probability that a Player 1 turn does win is $1 - \left(\frac{35}{36}\right)^{24} \approx 0.491$, which is a little less than one half. The probability that a Player 2 turn does not win is $\left(\frac{5}{6}\right)^4$, so a Player 2 turn does win with a probability of $1 - \left(\frac{5}{6}\right)^4 \approx 0.518$, which is a little more than one half.

For a Presentation: The document **Dice Games Present.ftm** has the collections and measures built. Open the inspector for the Two Dice collection and ask students to explain the formulas for *Die_1* and *Die_2* before you make the scatter plot (step 3) and ask Q2. Similarly, ask students to explain the formula for *Double_Ace* before you ask Q4 and the formula for *Game_Won* before you ask Q5. As you collect measures, start with five and leave the animation on so students can see that the measures are coming from the Two Dice collection. Then turn the animation off and collect 100 measures. Each time before you (or the student running the computer) click **Rerandomize** or **Collect More Measures,** ask students what they think the result will be. To explore the simulation for Player 2, scroll to the right. Extend the discussion to theoretical probabilities and include the information in the Extension to help students gain a complete understanding of the player most likely to win.

Q1 Student guesses will vary.

INVESTIGATE

Q2 A dot will be at the point (1, 1) if, and only if, the turn wins.

Q3 Because winning and not winning have almost the same chance, answers will vary. About 42% of the students will win more often, 33% will not win more often, and 25% will break even.

Q4 This is a bar graph. If there are any "true," then the turn wins. The percentages should break along the same lines as those in Q3.

Q5 The turn wins the game if *Game_Won* is true. Percentages among students should be about the same as in Q3 and Q4.

Q6 This result will probably also suggest an approximately even chance of wins.

Q7 Although there are more elaborate ways to respond to this question, most students will think of collecting a much larger number of measures. Doing so will probably convince a little more than half the students that the probability is less than 0.5. Because the inspector of Measures from Two Dice is not set to replace existing cases, clicking **Collect More Measures** will increase the size of the sample.

Q8 If they collect enough measures, students may find that a Player 2 turn is more likely to win than not. You might point out that this difference in likelihoods does not imply that Player 2 is more likely to win the game. By virtue of going first, Player 1 has an advantage. (In fact, if students work on the Extension, they will need to pursue infinite geometric series and find that Player 1 has a probability of almost 65% of winning.)

EXPLORE MORE

1. Student simulations should approximate the theoretical probability: The probability of a double is $\frac{6}{36}$, or $\frac{1}{6}$. The probability of not getting any doubles in n games is $\left(\frac{5}{6}\right)^n$. Solving $\left(\frac{5}{6}\right)^n = 0.5$ gives $n = \frac{\log(0.5)}{\log\left(\frac{5}{6}\right)} \approx 3.8$. It should take four rolls of the dice to make a fair game.

2. A first guess might use the fact that 4:6 = 24:36 = 144:216, which indicates 144 rolls. The situation is not linear, though. The probability of not getting a triple ace in n rolls of three dice is $\left(\frac{215}{216}\right)^n$. The value of n that makes the probability of not winning about the same as the others, 0.5, is the solution to the equation $\left(\frac{215}{216}\right)^n = 0.5$, which is $n = \frac{\log(0.5)}{\log\left(\frac{215}{216}\right)} \approx 149.37$. So, almost 150 rolls of three dice are needed until the probability of a triple ace is about 0.5.

EXTENSION

Which of the two original players has the greater probability of winning the game?

Answer: Player 1 wins on the first turn, or (if Player 2 also loses on his first try) on the third turn, or on the fifth turn and so on. The sum of that geometric series is

$$\textit{Player 1 wins}\frac{1}{1-(1-\textit{Player 1 wins})(1-\textit{Player 2 wins})} =$$
$$0.491\frac{1}{1-(0.509)(0.482)} \approx 0.65$$

Probability—Euchre Deck

A Euchre (pronounced "yoo-ker") card deck contains 12 red cards and 12 black cards. In one game played with this deck, each player begins with a five-card hand. How likely is a hand of all one color?

Q1 What would you guess the probability is?

Q2 Do you think the probability of a one-color hand would be higher or lower if you are dealt one card from each of five decks instead of all cards from the same deck?

INVESTIGATE

If you're simulating the flip of a coin or the roll of a die in Fathom, you usually set up a single case, with data generated randomly. If you're selecting randomly from a deck of cards, as in this exploration, you set up the data without randomness and then choose from it randomly. Fathom allows random choices through what is called *sampling*.

1. Open the document **Euchre Deck.ftm.** The collection represents the 24 cards in a Euchre deck. To pick cards at random from the deck, select the collection and then choose **Collection | Sample Cases.** A new collection is formed—Sample of Deck. If you enlarge the window to examine the new collection, you'll see that Fathom has chosen ten cards at random.

2. You want to simulate hands of five cards. So rename Sample of Deck to Hand and open this collection's inspector. Set the size of the hand to five cases and uncheck **With replacement.** (Unchecking **With replacement** prevents Fathom from choosing any card more than once for a hand.)

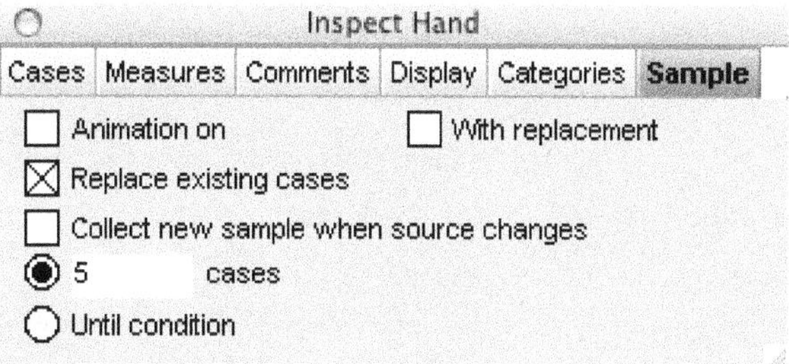

3. Click on **Sample More Cases** to deal another hand. Are the cards all the same color? Click on **Sample More Cases** until you have seen ten hands. Count how many hands have cards of only one color.

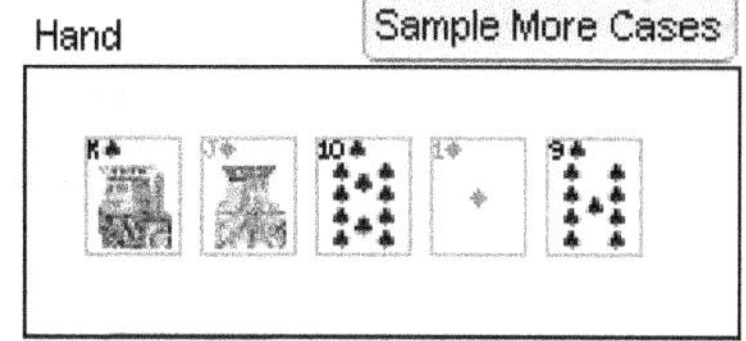

Q3 What percent of those ten hands had only one color? What would you estimate is the probability that a Euchre hand will be all red or all black?

You can get Fathom to do the counting for you if you tell it what to count by creating a measure.

4. Open the Hand inspector. Under the **Measures** tab, create a new measure with a name like *One_Color*. The formula you enter for this measure will use the variable *Red*, which is 0 for each black card and 1 for each red card:

   (sum(red) = 5) or (sum(red) = 0)

5. Continue to click **Sample More Cases,** counting the clicks, until *One_Color* has value true.

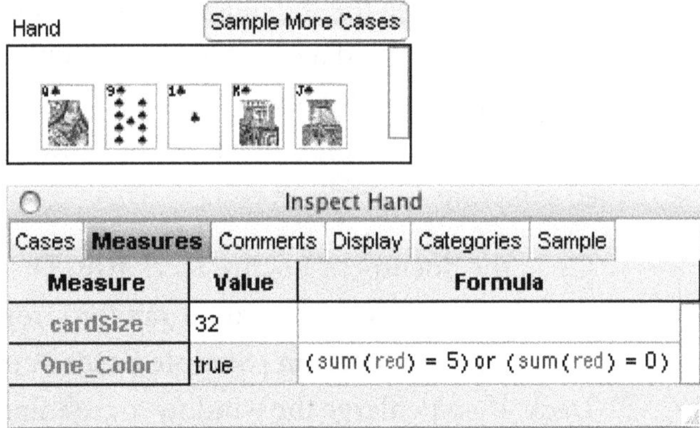

Q4 Now what would you guess about the probability of getting a hand that is all red or all black?

Animation helps show that the data cases in the collection Measures from Hand come from Hand, but it will make the sampling very slow. Uncheck **Animation on.**

6. Because this event might have a small probability, you will collect a large number of measures. Select the Hand collection and choose **Collection | Collect Measures.** Another collection will appear, called Measures from Hand. Open this new collection's inspector and set it to collect results on 1000 hands.

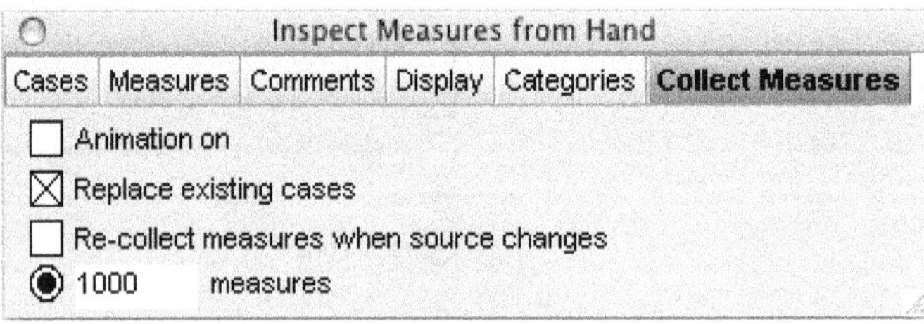

The attribute can be dragged from under the **Cases** tab in the inspector.

7. Click on **Collect More Measures** so that 1000 hands are simulated. To see how many of the 1000 hands had only one color, drag a summary table from the shelf and add the *One_Color* attribute from the **Cases** panel of this measures collection.

Q5 What do the results suggest about the probability of a hand of one color?

8. You've been considering five-card hands dealt from a single deck. With only one change, you can simulate dealing the five cards from different decks, thus allowing you to get the same card more than once. All you have to do is check **With replacement** in the inspector for Hand.

Q6 What is your estimate of the probability of a one-color hand if the cards come from different decks? Is it higher than, lower than, or equal to the probability if the cards come from a single deck? Explain why you think the probabilities have this relationship.

Q7 In probability, an event like a five-card hand might be considered five single events. These smaller events are either *independent* or *dependent.* Guess at the meaning of these words in the context of this problem.

EXPLORE MORE

A very poor hand in Euchre is a hand that has only nines and tens in it. Change your simulation to find the probabilities of this type of hand, dealt both from a single deck and from five decks. Is the relationship between the probabilities the same as the one you reported in Q6? (You will probably want to measure more than 1000 hands to find these probabilities.)

Objective: Students will create simulations to answer probability questions. They will contrast sampling with and without replacement and will gain an understanding of the difference between independent and dependent events.

Student Audience: Algebra 1, Algebra 2

Activity Time: 30–40 minutes

Setting: Paired/Individual Activity (use **Euchre Deck.ftm**) or Whole-Class Presentation (use **Euchre Deck Present.ftm**)

Mathematics Prerequisites: Students can calculate probability from a ratio.

Fathom Prerequisites: Students can sample cases from a collection, create measures and enter formulas, collect measures and use collection inspectors, and use summary tables.

Notes: As you introduce the activity or talk with students as they work, help them understand two powerful tools of Fathom—the ability to collect samples and the ability to collect measures from those samples. Collecting samples involves selecting cases at random from the original collection to form a new collection. Here, the original collection is the 24 cards in the deck; the sample collection is five cards drawn from the deck.

To keep track of the number of hands of one color, set up a measure that defines a hand of one color, then collect that measure as several hands are dealt. When Fathom collects measures, it re-collects the sample, calculates the formula created for each measure (determines whether the hand is one color), and builds a new collection with the results. The attributes in this new collection are the measures—here *One_Color*—and each case represents a measure taken from a different five-card hand. If you take measures from a large number of hands, you'll get close to the theoretical percentage of one-color hands in the long run.

After students finish the activity, you might combine simulation results in a table of the results for the entire class. How do the results compare? If a pair completed Explore More, ask them to share their results.

In this case, sampling without replacement creates dependent events. The theoretical probability of a hand of one color using a deck of 24 cards with 12 cards in each color can be calculated:

$$\frac{\binom{12}{5}}{\binom{24}{5}} = \frac{\frac{12!}{5!7!}}{\frac{24!}{5!19!}} = \frac{12!19!}{24!7!} = \frac{3}{161} \approx 0.0186\text{, or 19 in 1000}$$

But the case of sampling with replacement creates independent events. Here the theoretical probability is equal to that of taking one card from each of five decks.

$$\frac{12^5}{24^5} = \left(\frac{1}{2}\right)^5 = \frac{1}{32} \approx 0.0312\text{, or 31 in 1000}$$

For a Presentation: The presentation document **Euchre Deck Present.ftm** starts with the Hand collection ready to sample. Start by sampling more cases and asking students how many hands of one color they think will result from sampling ten hands. As you continue with step 3, ask students what they expect will happen. Ask several students to share answers to Q5–Q7.

Q1 Answers will vary. Some students may say the probability is $\frac{1}{2}$, because half the cards are each color. This is not the time to correct these misconceptions.

Q2 Answers will vary. This is not a point at which you need to clarify these ideas.

INVESTIGATE

Q3 Students are very likely to have seen zero hands of the same color, so the most common answer will be that the probability is small.

5. This step may take a while. The expected value is an additional 53 clicks to find a true. If any students have not seen true by the time they get 100 clicks, tell them to move on.

Q4 Students will likely divide 1 by the number of clicks, giving a value somewhere between 0.01 and 0.1, in most cases.

6. Even with animation off, taking 1000 measures will take about a minute.

7. 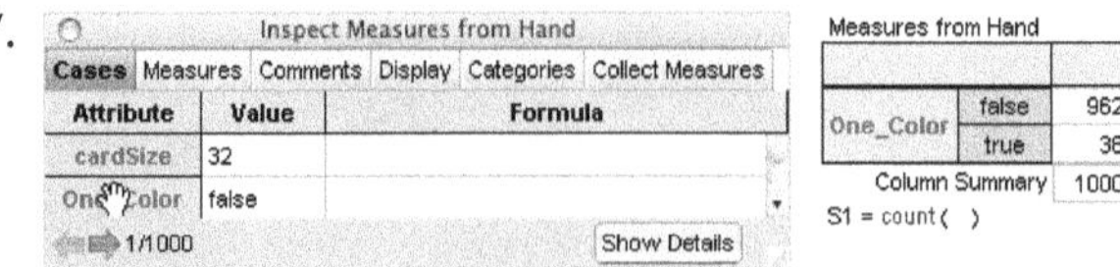

Q5 The theoretical probability is approximately 0.019. About 95% of the values that students get will lie between 0.010 and 0.027.

Q6 The theoretical probability is about 0.031. You can expect 95% of the values to be between 0.020 and 0.042. Fewer than 1 in 20 students will find this event less likely than the first event. Ask them to sample another 1000 hands and see if their opinion changes.

Students may note that when dealing from one deck, if they've already received one or more cards in the same color, then it will be less likely that the next card will be the same, because there are fewer cards left of that color. In contrast, when dealing from different decks, the colors of the previous cards dealt have no bearing on the next card, which is still equally likely to be either color.

Q7 The events from the one-deck deal are dependent, because the probability changes depending on what cards have been dealt. The events from the five-deck deal are independent, because the probability for the last cards is independent of what the first cards are.

EXPLORE MORE

These probabilities can be found by creating a measure *Nine_Ten* in the Hand collection that is equal to count(number = 9) + count(number = 10) and looking at a histogram to determine the number of cases where the total count is five. Students may change this into (count(number = 9) + count(number = 10)) = 5 and then count the occurrences of true that result. The theoretical probability is 0.0013 from a single deck and three times that amount—0.0041—from multiple decks.

Simulation—Stick or Switch

Imagine a game with three doors, numbered 1, 2, and 3. Behind one of the doors is a fantastic prize, like a car, and behind the other two are consolation prizes, like goats. After you choose a door, but before the door is opened, the game host opens another door to expose a consolation prize and asks whether you would like to switch your guess.

Q1 When you first select a door, what is the probability of choosing the door with the prize?

Q2 Do you think you have a better probability of winning if you switch doors after a losing door is opened or if you stick to your original guess? Explain.

INVESTIGATE

A Fathom simulation can help you answer Q1 and Q2. To set up the simulation, you'll take the point of view of the game show host, who knows which prize is behind each door.

1. In a new Fathom document, create a collection with attributes such as *First_Guess* and *Winning_Door*. Both the winning door and the contestant's first guess will be random, so give each attribute the formula randomInteger(1,3).

You might name your collection Game.

2. To record whether the first guess is a winning guess, create a new attribute, *Win_By_Sticking*, with the formula First_Guess = Winning_Door. This attribute will have value true or false. Highlight the collection and choose **Collection | New Cases;** add 100 new cases. Drag down a new graph and graph *Win_By_Sticking* on a bar chart.

You can select the bar and read the value from the status window at the bottom left corner of the Fathom window.

Q3 In how many of these 100 games does *First_Guess* equal *Winning_Door*? How well does this number agree with your answer to Q1?

3. If you rerandomize, the selected cases will distribute differently in the two bars. To keep track of the number of cases in each category, you can create a measure. In the inspector of the Game collection, create a measure called *Sticking_Probability* with the formula shown here.

Open the inspector for the collection and click on the **Measures** tab, then click on the word *<new>*.

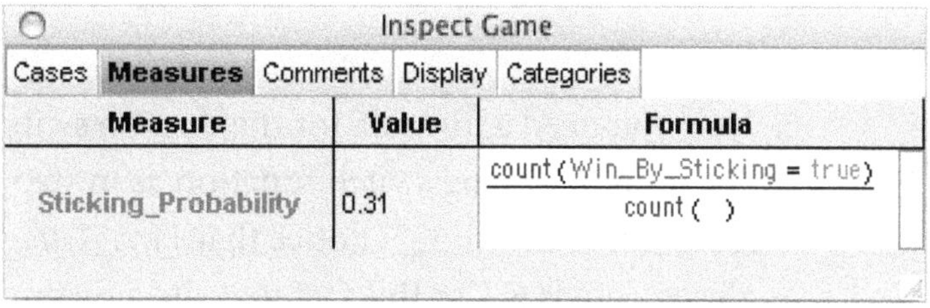

Inspect Game

Cases | **Measures** | Comments | Display | Categories

Measure	Value	Formula
Sticking_Probability	0.31	$\frac{\text{count(Win\_By\_Sticking = true)}}{\text{count( )}}$

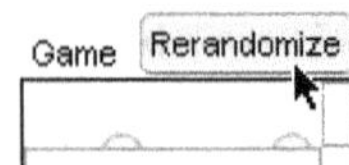

Q4 Rerandomize a few times and watch the value of *Sticking_Probability* change. What are the maximum and minimum values you observe when rerandomizing repeatedly?

Q5 You now need to create a function that will simulate the opening of a door that is not the door you selected and not the door with the prize. You'll also keep track of which door you'd switch to if you did switch. To help you organize, make and complete, by hand, a table like this for all nine possibilities of *First_Guess* and *Winning_Door.*

First_Guess	*Winning_Door*	*Shown_Door*	*Switch_Door*
1	1	2	3
		3	2
	2	3	2
	3	2	
2	1		

4. Open the sample document **Three Doors.ftm.** Show the collection's inspector and look at some of the cases.

Q6 Use a few examples to determine what numbers could result from the formula $10 \cdot \textit{First\_Guess} + \textit{Winning\_Door}$.

To insert a new case in the formula editor, put the cursor on the last full case and press Ctrl+Enter (Win) or Option+Return (Mac). The randomPick function picks randomly from a list of choices.

5. The switch formula for *Shown_Door* has been completed only partially. Use the table you created to add the missing test values to the formula for *Shown_Door.* Examine a case of each of the nine types to make sure the formula is working properly.

6. Now you need a formula for the door to switch to, *Switch_Door.* You could create another long switch function as in step 5, but there's an easier way. In the table you made for Q5, notice that *First_Guess* + *Shown_Door* + *Switch_Door* always equals 6. Use this fact to write a formula for *Switch_Door.*

Q7 What formula did you use?

7. Create a *Win_by_Switching* attribute and its formula the way you did in step 2.

Q8 What is your formula for the attribute *Win_by_Switching*?

8. Create a new measure called *Switching_Probability* with a formula similar to the one used in step 3. With the **Measures** tab open, rerandomize to replay the 100 games.

Q9 What is your formula?

Q10 What is the sum of *Switching_Probability* and *Sticking_Probability* every time the collection is rerandomized? Why? Which strategy, sticking or switching, has the better probability? By how much?

EXPLORE MORE

1. Suppose there are four doors but still only one prize, and the game is played the same way. Should the contestant switch to one of the two other unopened doors? What would the probability of winning by sticking and switching be now? Design a Fathom simulation to support or contradict your belief. (*Note:* You will have to make a long switch function in step 6, because there is no clever shortcut.)
2. You have three coins in your pocket. One has two heads, one has two tails, and one is a normal coin. You reach in and pull out a coin at random. What is the probability that when you flip the coin it lands heads? If it does land heads, then what is the probability that there is a head on the other side of the coin? Design a Fathom simulation to support or contradict your belief.

Simulation—Stick or Switch

Objective: Students will make sense of probability, will be introduced to conditional probability, and will create probability simulations.

Student Audience: Algebra 1, Algebra 2

Activity Time: 25–40 minutes

Setting: Paired/Individual Activity (use **Three Doors.ftm**)

Mathematics Prerequisites: Students can calculate probability from a ratio.

Fathom Prerequisites: Students can create a collection, add attributes and formulas, create measures and enter formulas, rerandomize a collection, and create a bar graph.

Background: This problem is based on a scenario taken from the game show "Let's Make a Deal," which ran from 1963 until 1977. The host, Monty Hall, presented the contestant with three doors, numbered 1, 2, and 3, one of which had a fantastic prize behind it. People have speculated on the best strategy to follow if, after the contestant chose a door but before the door was opened, Monty had another door opened to expose a consolation prize. Would it be better to switch the choice to the other unopened door? A column of "Ask Marilyn" in *Parade* magazine asserting that the contestant would be better off to switch induced thousands of letters of dispute, including some from professional mathematicians. A correctly constructed simulation can show that Marilyn was right.

Notes: Fathom's formula editor is tricky to use in completing step 5; essential help is given in the margin note on the activity page. If students wonder what *else* means at the bottom of the switch statement, indicate that a switch statement always ends with *else* to tell the computer what to do if none of the other statements hold. It is often used for the last of several alternatives. You might ask what could be included in the *else* here. Students who include (33) in the body of the switch and leave the *else* blank will still be able to complete the activity.

The Explore More activities give students the opportunity to apply what they just learned about simulations. You might ask students who complete either of the Explore More activities to share their method and result with the class.

The counterintuitive solution leads naturally to the question "Why?" You might ask students to give explanations that make sense to them. Explanations that satisfy some people don't satisfy others. One explanation is that the probability was $\frac{1}{3}$ that the prize was behind the chosen door and that probability does not change by opening another door. Therefore, the probability of winning by switching doors is $\frac{2}{3}$.

Q1 Answers will vary.

Q2 Answers will vary. Students may argue that because the probabilities that the prize was behind any of the three doors were the same, that's still true about the remaining closed doors. Or they might argue that because you now know that the prize is behind one of two doors, each door now has an equal chance of having the prize. So neither sticking nor switching carries a higher probability.

INVESTIGATE

Q3 The two values, *First_Guess* and *Winning_Door,* will be the same in about 33 cases, or $\frac{1}{3}$ of the 100.

Q4 Most values will stay around 33%. If you see values higher than 42%, you can suspect an error in the simulation.

Q5

First_Guess	*Winning_Door*	*Shown_Door*	*Switch_Door*
1	1	2	3
		3	2
	2	3	2
	3	2	3
2	1	3	1
	2	1	3
		3	1
	3	1	3
3	1	2	1
	2	1	2
	3	1	2
		2	1

Q6 Calculating 10 • *First_Guess* + *Winning_Door* is an arithmetic device to turn a pair of one-digit numbers into a single two-digit number.

5.

Shown_Door	3	switch (10First_Guess + Winning_Door)	(11) : randomPick (2, 3) (12) : 3 (13) : 2 (21) : 3 (22) : randomPick (1, 3) (23) : 1 (31) : 2 (32) : 1 else : randomPick (1, 2)

Q7 *Switch_Door* = 6 − *First_Guess* − *Shown_Door*

Q8 *Switch_Door* = *Winning_Door*

Q9 $\dfrac{\text{count}(\textit{Win\_By\_Switching} - \textit{true})}{\text{count}()}$

Q10 The sum of *Switching_Probability* and *Sticking_Probability* will always be 1 because there are no other doors that could have the big prize.

You should always switch, because it doubles the probability of choosing the big prize.

EXPLORE MORE

1. The theoretical probabilities here are 0.25 to win by sticking and 0.375 to win by switching. So the advantage is only slightly increased by switching.

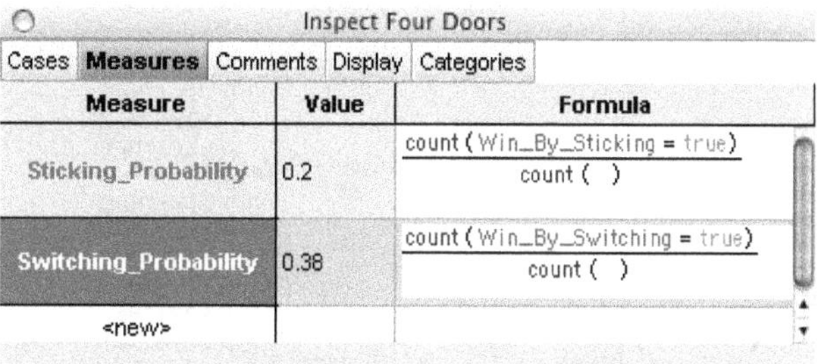

Inspect Four Doors

Cases | Measures | Comments | Display | Categories

Measure	Value	Formula
Sticking_Probability	0.2	count (Win_By_Sticking = true) / count ( )
Switching_Probability	0.38	count (Win_By_Switching = true) / count ( )
<new>		

EXPLORE MORE 1

Inspect Four Doors

Cases | Measures | Comments | Display | Categories

Attribute	Value	Formula
First_Guess	2	randomInteger (1, 4)
Winning_Door	2	randomInteger (1, 4)
Win_By_Sticking	true	First_Guess = Winning_Door
Shown_Door	4	switch (10First_Guess + Winning_Door) { (11) : randomPick (2, 3, 4) (12) : randomPick (3, 4) (13) : randomPick (2, 4) (14) : randomPick (2, 3) (21) : randomPick (3, 4) (22) : randomPick (1, 3, 4) (23) : randomPick (1, 4) (24) : randomPick (1, 3) (31) : randomPick (2, 4) (32) : randomPick (1, 4) (33) : randomPick (1, 2, 4) (34) : randomPick (1, 2) (41) : randomPick (2, 3) (42) : randomPick (1, 3) (43) : randomPick (1, 2) else : randomPick (1, 2, 3)
Switch_Door	1	switch (10First_Guess + Shown_Door) { (12) : randomPick (3, 4) (13) : randomPick (2, 4) (14) : randomPick (2, 3) (21) : randomPick (3, 4) (23) : randomPick (1, 4) (24) : randomPick (1, 3) (31) : randomPick (2, 4) (32) : randomPick (1, 4) (34) : randomPick (1, 2) (41) : randomPick (2, 3) (42) : randomPick (1, 3) else : randomPick (1, 2)
Win_By_Switching	false	Switch_Door = Winning_Door
<new>		

7/100 Show Details

2. There is an even chance that you will see heads or tails on the flip because the six sides are equally likely—three heads and three tails. But if the flip should show one of the three possible heads, then the probability is that the other side is also a head.

Inspect Three Coins

Cases | Measures | Comments | Display | Categories

Attribute	Value	Formula
Coin	3	randomInteger(1, 3)
Side	2	randomInteger(1, 2)
Top	Tail	switch(10coin + Side) { (11) : "Head" (12) : "Head" (21) : "Tail" (22) : "Tail" (31) : "Head" else : "Tail"
Bottom	Head	switch(10Coin + Side) { (11) : "Head" (12) : "Head" (21) : "Tail" (22) : "Tail" (31) : "Tail" else : "Head"
<new>		

2/100 Show Details

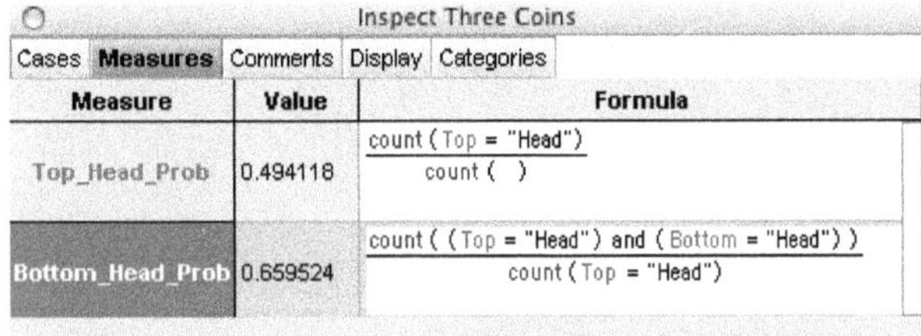

Inspect Three Coins

Cases | Measures | Comments | Display | Categories

Measure	Value	Formula
Top_Head_Prob	0.494118	count(Top = "Head") / count( )
Bottom_Head_Prob	0.659524	count((Top = "Head") and (Bottom = "Head")) / count(Top = "Head")

Notice that in the formula for *Bottom_Head_Prob*, the numerator is not simply count(bottom = "Head"), nor is the denominator simply count( ). This is because we are looking for the probability that the bottom is heads *if* the top is heads. That is, we are only counting the cases (both in the numerator and in the denominator) where we already know that the top is heads.

EXTENSION

Students might research the background of the problem, variously known as the "Monty Hall" problem or the "Let's Make a Deal" problem, on Web sites such as www.letsmakeadeal.com and www.marilynvossavant.com.

Geometry by Probability—Monte Carlo Method

Did you know that you can use probability to estimate the area of an irregular two-dimensional shape? Before looking at a less regular shape, here is an example that is simple enough for you to check your estimate by calculation.

Q1 Estimate what fraction of the square is shaded.

Q2 If each side of the square measures 5 cm, what does your estimate say is the area of the shaded region?

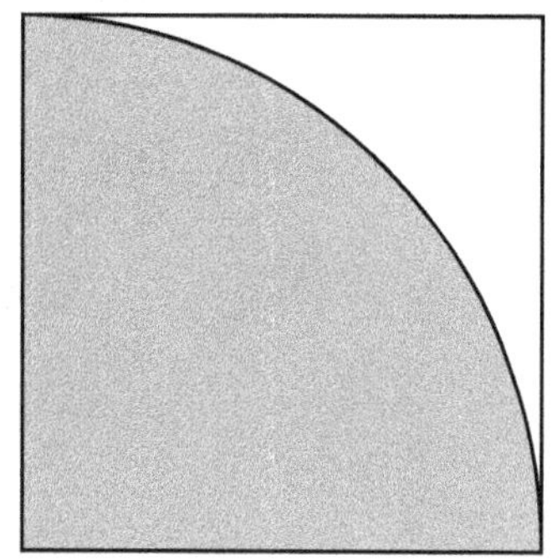

INVESTIGATE

You can use probability to estimate because the probability that a random point in the square will lie in the shaded region equals the ratio of the areas of shaded to square. You can use Fathom to choose random points in the square by their coordinates.

You might call your collection Points in Square.

1. Create a collection with two attributes for coordinates x and y and give each attribute the formula random(5). This formula makes each coordinate a random number between 0 and 5. Choose **Collection | New Cases** to add 1000 new cases. Graph y against x on a scatter plot.

Q3 Describe the scatter plot.

To mark the shaded region, you first note that the points inside are all less than 5 units from the origin.

Q4 What inequality describes the coordinates of the shaded points?

2. Create a new attribute, named something like *Shaded*, that will take on the value *true* if the point (x, y) is inside the shaded area. Use the inequality from Q4 in its formula.

When you drop the variable on the horizontal axis, it will default to a bar chart. Change the plot type in the upper corner to Ribbon Chart.

3. Create a new ribbon chart of *Shaded*.

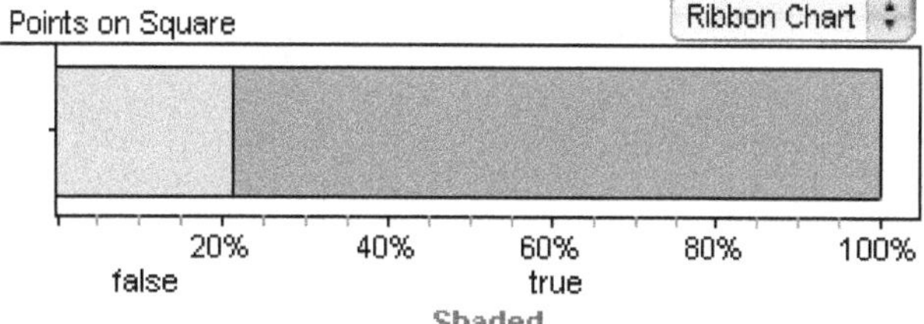

Q5 How can you mark the points that are to be shaded on the scatter plot?

Q6 What percentage of the ribbon chart is true? How close is this to your prediction in Q1?

You'll need to take two steps of 5,000 each.

4. Add 10,000 new cases to the collection.

Q7 Estimate the area of the shaded region.

Click on the white space to the right of the ribbon if you want to deselect the points.

Q8 The shaded region is one quarter of a circle. Use the estimate from Q7 to estimate the area of the full circle with radius 5 cm.

Q9 The conventional formula for the area of a circle has the form $A = kr^2$, where r is the radius. Put in your values for A and r and find the corresponding value of k. What can you say about this method for finding area?

5. Repeat the investigation, adding or changing attributes to estimate the area above, or inside, the parabola $y = (x - 2.5)^2$, where y is between 0 and 5.

Q10 What is your estimate of the area inside the parabola? Justify your ideas.

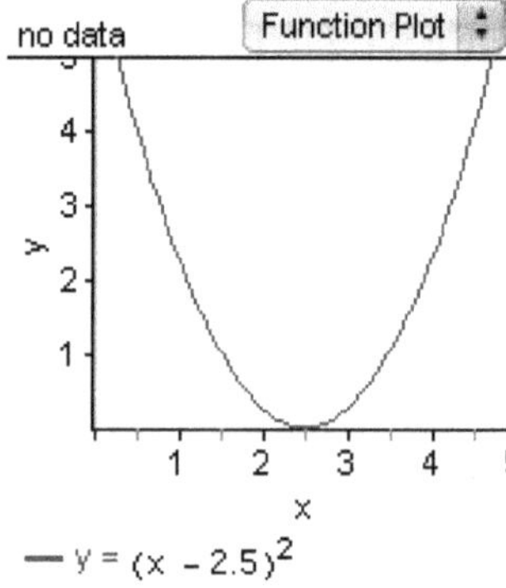

EXPLORE MORE

1. Look at ratios and areas for other quarter-circles in squares. Start with a slider named *Size* and use it in place of the 5 in the attribute formulas. Set the slider to different values and report your findings on the value of k in $A = kr^2$.

2. Look for a pattern in the areas of the parabola given in step 5. Start with a slider named *Vertical* that is restricted to integer values. Change the formula for the y-attribute to random(*Vertical*) and find the area for heights of 1, 2, 3, and 4. Along with the value you found in Q10 for a height of 5, can you discover any patterns in these numbers? Try using sliders a and b to find a line of fit for $Area = a \cdot Height^b$.

Geometry by Probability—Monte Carlo Method — Activity Notes

Objective: Students will use a probability simulation to estimate area.

Student Audience: Algebra 1, Algebra 2

Activity Time: 25–35 minutes

Setting: Paired/Individual Activity

Mathematics Prerequisites: Students can use the distance formula, calculate the area of a square, and work with ratios.

Fathom Prerequisites: Students can create a new collection, add attributes with formulas, and create a scatter plot and a ribbon graph.

Background: The term *Monte Carlo Method* refers to a method that solves a problem by generating suitable random numbers and observing the fraction of those numbers that obey particular properties—in this case, the fractions that are within the shaded area. The method was first given a name in 1946 when a mathematician named it for a city known for gambling.

Notes: As you facilitate student work, you may find students struggling with step 2. To help them see how to use the distance formula, point to one point and ask how far it is from the origin, then ask them to generalize to a point (x, y). How far is it from the origin? You might remind students working on Q5 that the percentages can be read off the status bar. Not all students will recognize π as the constant in Q9. In a class discussion after the activity, you might ask for area estimates (answers to Q8) from several students. Finding the average and then dividing by 4 might yield a constant close enough to 3.14 to be recognized as π.

Students who already know the formula for the area of a circle should be convinced by the first part of the activity that you can use random numbers to estimate area. They will then be convinced of the area of the parabola. Students who do not know the formula for the area of a circle may discover it.

This group will benefit especially from Explore More 1, where they look at circles with different radii. Those who already know the circle formula will benefit more from Explore More 2, where they look for patterns and try to generalize the formula for parabola areas.

Q1 Guesses will vary. The actual fraction of shaded area in the quarter-circle is $\frac{\pi}{4} \approx 0.785$.

Q2 A good estimate of the area is about 20 square units, or 80% of the square. The actual area is about 19.6 square units.

INVESTIGATE

Q3 The points should fill the square from (0, 0) to (5, 5).

Q4 By the distance formula or the Pythagorean theorem, $\sqrt{x^2 + y^2} \leq 5$. Answers using a strict inequality ($<$) are also acceptable.

Q5 Select the region of the ribbon chart corresponding to *Shaded = true.*

Q6 The percentage will likely (95% of the time) be between 76 and 81.

Q7 With this many cases, the percentage will likely (95% of the time) be between 0.778 and 0.793. Rerandomizing a few times and averaging the results will give the value 0.785, leading to an area of 19.625 square units.

Q8 $19.625 \cdot 4 = 78.5$ square units

Q9 k is about 3.14. Point out that this multiplier has the name, π, *pi;* as needed, state the formula for the area of a circle as $A = \pi r^2$.

Q10 The formula $y \geq (x - 2.5)^2$ for *Shaded* leads to a ratio of about $0.\overline{6}$ and an area of about 15 square units.

EXPLORE MORE

1. For various values of the radius, or *Size*, k will be close to π.

2 Experimental values of (*Vertical, Area*) will be close to the theoretical values of about (1, 1.33333), (2, 3.77124), (3, 6.9282), (4, 10.66667), and (5, 14.9071). The values are increasing faster than in a linear relationship but not so fast as in a quadratic relationship. The best fit model (actually an exact fit to the exact values) is $Area = \frac{4}{3}Height^{\frac{3}{2}}$, or $y = 1.33x^{1.5}$.

EXTENSION

Parallel lines are drawn 10 cm apart, filling a sheet of paper. A 10 cm needle is then dropped on the paper. What is the probability that it will cross one of the lines?

Answer: It can be proved using trigonometry, geometric probability, and calculus that the probability will approach $\frac{2}{\pi}$ in the long run. Answers from large numbers of trials will be near 0.6366. Historically, the experiment was used to estimate the value of π as equal to 2 divided by the experimental probability.

www.ingramcontent.com/pod-product-compliance
Lightning Source LLC
Chambersburg PA
CBHW061406210326
41598CB00035B/6116

* 9 7 8 1 5 5 9 5 3 8 0 1 5 *